THE SUICIDAL PERSON

THE SUICIDAL PERSON

A NEW LOOK AT A HUMAN PHENOMENON

KONRAD MICHEL

Columbia University Press   *New York*

Columbia University Press
Publishers Since 1893
New York   Chichester, West Sussex
cup.columbia.edu

Copyright © 2023 Konrad Michel
All rights reserved

Library of Congress Cataloging-in-Publication Data
Names: Michel, Konrad, author.
Title: The suicidal person : a new look at a human
phenomenon / Konrad Michel.
Description: New York : Columbia University Press, [2023] |
Includes bibliographical references and index.
Identifiers: LCCN 2023002747 (print) | LCCN 2023002748 (ebook) |
ISBN 9780231205306 (hardback) | ISBN 9780231555944 (ebook)
Subjects: LCSH: Suicide—Psychological aspects. | Suicidal behavior. |
Suicide—Treatment. | Stress (Psychology)
Classification: LCC HV6545 .M445 2023 (print) | LCC HV6545 (ebook) |
DDC 362.28—dc23/eng/20230513
LC record available at https://lccn.loc.gov/2023002747
LC ebook record available at https://lccn.loc.gov/2023002748

Cover design: Mary Ann Smith

Dedicated to

Alex

Terry Maltsberger

My patients whose lives I couldn't save

CONTENTS

Acknowledgments ix
Introduction xi

1 Losing a Patient to Suicide and What It Means for a Young Doctor 1

2 First Lessons in Reducing Suicide 17

3 Emotional Stress Affects Brain Function 37

4 The Brain and Suicide 57

5 Problems of Communication in Medical Consultation 81

Box: Theories to Explain Suicide 93

6 Suicide Is Not an Illness 103

7 The Fragile Sense of Who We Are 123

8 Personal Vulnerabilities and Suicide 139

9 A Think Tank of Concerned Therapists 161

10 Translating Acquired Knowledge Into a New Therapy 181

11 Now What Does This All Mean for Suicide Prevention? 207

12 A Special Concern: Young People   223

13 For Health Professionals: It's About the
         Person in the Patient   249

Appendix 1. Suicide Is Not a Rational Action:
ASSIP Handout for Patients   275
Appendix 2. Questionnaire: Learning Objectives   281
Appendix 3. What to Do If You Are Concerned About Someone   287
Appendix 4. Resources   291
Notes   297
Bibliography   325
Index   333

ACKNOWLEDGMENTS

Thanks to Raj Rathod and Anthony Parker, who helped me develop the skills needed for good clinical practice.

Thanks to the Royal College of Psychiatrists and its training scheme, which allowed me to be proud of my identity as a psychiatrist.

Thanks to Michael Gelder and Tamara Kolakowska, who introduced me to research.

Thanks to Edgar Heim, who taught me the art of psychotherapy.

Special thanks to my friend and most wonderful colleague Ladislav Valach, the person who changed my views on suicide and the approach to the suicidal person.

Thanks to David Jobes, John Maltsberger, Israel Orbach, Mark Goldblatt, David Rudd, Gregory Brown, Marsha Linehan, Michael Bostwick, Lisa Firestone, and all the other dedicated clinicians and researchers who generously shared their ideas and experiences at the Aeschi Conferences.

But most of all, I want to thank the hundreds of patients and their families that readily shared their stories with me. They made me feel extremely privileged by trusting me with their most intimate inner experiences. Without them, the present book would have remained a sterile intellectual exercise.

My very, very special thanks to Wendy Keller (Keller Media) for her incredible engagement in preparing my manuscript for submission to the publisher and her strong belief in my book project.

I also want to thank the two wonderful—anonymous—reviewers of my manuscript who convinced Columbia University Press to go ahead with my book project.

INTRODUCTION

Usually we can rely on our rational mind and past experiences to make the right decisions.

Some decisions turn out to be wrong, but we are able to correct the situation.

Some decisions cannot be corrected. Some are deadly.

Suicide is one of them.

The first suicide of a patient of mine happened in my early training years. Gil was married and a mother of two small children, and her death was an experience that had a long-lasting effect on my professional development. For nearly fifty years and with the help of hundreds of patients I have been trying to understand one of the most tragic and enigmatic human behaviors.

Why do nearly a million people around the world die by suicide each year? How is it possible that an emotional crisis can threaten our own physical existence? This book invites you to take a new look at a human phenomenon that is difficult to understand.

Many theoretical models try to explain suicide. The prevailing medical model is focused on risk factors. Health professionals are taught that when faced with a suicidal patient, they must first of all diagnose and treat disorders of mental health such as depression, psychosis, or substance use. Yet medical risk factors are statistical data. They cannot do justice to the individual suicidal person with a personal suicidal logic and a unique life story. Current concepts of suicidality do not meet the needs of people at risk. If we continue to see suicide through the medical lens—as a pathological condition—and as a consequence of a mental illness, we will not get any further in reducing suicide worldwide. What we need are concepts to understand the suicidal person that go beyond the medical model of suicide.

My main professional identity is that of a clinical psychiatrist and psychotherapist. My second professional identity came to be that of a researcher. This combination makes me a rather rare specimen in the world of suicide prevention, in that I believe that the ultimate objective in research must be to reduce the number of tragic and unnecessary deaths by suicide. At the closing session of an international suicidology congress I made a list of the titles in the book of abstracts and presented the results: the majority of contributions focused on the epidemiology of suicide, medical and psychosocial risk factors, and the neurobiology of suicide. Nine percent dealt with treatments of suicidality (including medication) and just 4 percent with psychotherapy for suicidal patients.

As a trained psychotherapist I have learned that an effective therapy must be built on a meaningful therapeutic relationship between patient and therapist. Yet I have found that most concepts of suicidal behavior are not very helpful for therapy. This is where Ladislav Valach, my ingenious and dear colleague, came into the picture. As an expert in health psychology, Ladislav had a very different professional background from mine as a medically trained psychiatrist. He was the one who opened my eyes to see suicide not as a symptom of a psychiatric disorder but as an action for which people have their personal reasons.[1] In 1997, we published a first case study based on a suicide-as-action framework.[2] We then invited a group of international suicidology experts to discuss our new concept, a meeting that in the following years grew into regular international conferences, where new developments in understanding and treating suicidal patients were shared and discussed with some of the best suicide experts and researchers in the world.[3] This exciting professional exchange was an amazing boost in my search for an effective therapy approach for patients with a high suicide risk. In our psychiatric clinic, situated on the campus of a university general hospital, we had an ideal situation, with close links to the Emergency Department, where all the patients who had made a suicide attempt were admitted. In hundreds of interviews with suicidal patients I asked people to tell me their stories, and I learned to sit back and first of all be an empathic, interested, and attentive listener. I learned that each suicide story is unique and that a one-size-fits-all approach to these patients cannot work.[4] But I also found that when it came to the suicide action, which could have ended in death, most patients were at a loss when trying to explain their out-of-the-ordinary mental state. They often used descriptions such as "This was not me," "I was in a trancelike state," or "I was acting like in an autopilot mode."

This is when I got interested in the neurobiology of suicide. In a study using functional brain imaging we showed that the suicidal mode as an on/off phenomenon, as was known from cognitive therapy, is also a neurobiological mode. The results supported our theory that in the acute suicidal mental state, the function of the prefrontal cortex is drastically reduced, which means that rational thinking and decision making are largely abolished. I felt that our patients must know that high levels of emotional stress can dramatically change brain function. When we experience extreme psychological pain, the emotional brain will overrule the rational brain and pursue just one short-term goal: to end this unbearable mental state—and often life with it. For our patients who had survived a suicide attempt this was an important insight. They learned that to survive future emotional crises they had to know and be able to identify their early warning signs and use their safety strategies before it was too late—that is, before the emotional brain could take over and drive them into an act of self-harm they would not usually do when clear-minded.

Today I am convinced that the vast majority of people who die by suicide would regret their final act if they had an opportunity to reconsider. I heard from people who survived jumping from a bridge (and here in Bern we have several high bridges over a river) that at the very moment they jumped they regretted what they had just done. In my work with suicidal patients it became a major goal to develop therapeutic concepts that would have a long-term preventive effect for people who, after attempted suicide, clearly lived with an increased suicide risk. My colleague Ladislav encouraged me to develop a novel therapy concept, in which patients became active partners in therapy. We developed the technique of narrative interviewing, followed by a video playback of the narrative interview, a powerful therapy component to increase insight. In this form of self-confrontation, a patient said: "This is frightening! I don't want to get into this state of mind again where I'm not myself." We further optimized the patient's engagement with treatment with a collaborative homework task, a jointly developed written case formulation, and personal safety planning. The end product was the Attempted Suicide Short Intervention Program (ASSIP), a three-session therapy program that will be described in detail later in this book.

The suicide-as-action model, which we have developed and refined over a number of years, introduces a new, person-centered understanding of the suicidal mind. The model does not attempt to explain suicide but describes the psychological dynamics of the suicidal person. It is a model with a strong emphasis on the suicidal person's biography, their vulnerabilities and important personal needs and life goals. Every suicide story is unique. However, I do not claim that this model of understanding can do justice to *all* suicides. Suicidal behavior is far too diverse. A history of early trauma can have such a devastating effect on a person's life that the focus on impaired rational thinking in the acute suicidal moment seems inadequate. I am also aware that severe and chronic mental illness and opioid dependency due to chronic pain can be reasons for ending life by suicide. And it may be difficult to fit chronic suicidality and cultural factors into the model.

Nor do I want to discard the medical model. Medical conditions are suicide risk factors and clearly require adequate professional evaluation and treatment. What I advocate is a treatment approach that goes beyond the medical model. We must put the uniqueness of the suicidal person in the center. To make a therapy meaningful to the patient, therapists must find a way to relate to "the person in the patient."

This book describes my personal and professional development and will take the reader through my journey into the suicidal mind. It will include many case examples but also helpful material for people affected directly and indirectly by suicide. There will be a chapter for health professionals—because without the helping profession being open to a new approach to the suicidal patient, little will change. We need care providers who meet the needs of the suicidal person.

One of the major problems in suicide prevention is that some 50 percent of people who die by suicide do not seek help before their final decision. The suicide-as-illness model is not helpful for many people at risk, because of the association of suicide with mental health disorders and the related stigma. I believe that promoting a model that puts the suicidal person's dynamics into the center has the potential to change attitudes toward suicide in the general population. With this book I hope to reach as many people as possible, including those who know about the topic from their own experience, those who have lost someone to suicide, those who are concerned about someone, those who simply want to learn more about a human phenomenon that is so difficult to understand, and those who are interested in basic human issues such as the sense of identity, the effect of early experiences, the nature of emotional crises, emotion regulation, brain function, and the art of psychotherapy.

Worldwide, over 800,000 people die by suicide every year; over 47,000 in the United States; 53,000 in Western Europe; and 87,000 in Central and Eastern Europe, with China and India constituting 44.2 percent of global suicide deaths.[5] In the United States, after a decline of two years, the suicide rate increased again from 2020 to 2021, with the largest increase for males ages fifteen to twenty-four. The number of suicide attempts in the United States is estimated at 1.2 million per year.[6] A special concern is the increasing suicide rate among young people in Western countries.[7]

Far too many tragedies and far too many lives lost.

I have come to know the tragedy of suicide not only as a professional.

I lost our son Alex, aged twenty, to suicide in 2001.

I couldn't help my own son.

I hope this book will help others.

Konrad Michel, August 2022
Bern, Switzerland

THE SUICIDAL PERSON

1

LOSING A PATIENT TO SUICIDE AND WHAT IT MEANS FOR A YOUNG DOCTOR

One morning when I arrived at the hospital, the nursing staff waited for me, a young trainee in psychiatry, with serious faces. Gil, my patient, forty-two-years old, married, with two children, had thrown herself under a truck in front of Roffey Park Hospital.

I had been faced with death in my internship in general medicine, and I had learned to cope with young people dying in the emergency room. But this was different. I was her therapist. And I was responsible for her.

Roffey Park, county of Sussex, in the south of England, is an impressive Jacobean-style manor house situated in a vast and lush park full of giant rhododendrons. It was built by the Kaiser family, a business family who made their wealth as merchants. After World War II, Roffey Park became a rehabilitation center for army returnees suffering from what was called "war neurosis"—which today would be post-traumatic stress disorder, PTSD. It then became a National Health Service (NHS) psychiatric hospital, and later it was sold and turned into a private residential home.

2 ∞ LOSING A PATIENT AND WHAT IT MEANS

FIGURE 1.1. Roffey Park Hospital, the location of my first patient suicide.

In 1977 I came there for my specialist training in psychiatry, a field I had chosen after two years in internal medicine, where I felt that I spent far too much time interpreting X-rays, ECGs, lab results, etc., leaving me hardly any time to get to know patients as human beings. I had read several classical textbooks by English authors and, after having dealt with a lot of red tape, finally succeeded in getting a training job in this psychiatric service, which offered inpatient, outpatient, and day hospital facilities. There were huge differences between England and Switzerland in the practice of psychiatry and psychiatric training, and working in this unique hospital felt like being on a different planet. Roffey Park provided an amazingly comprehensive service. My tasks as a newcomer to psychiatry ranged from applying ECT (electroconvulsive therapy or "electroshock") to patients (who, after treatment was over, got a cup of tea and went home!), to the regular morning rounds with all patients, outpatient and day hospital sessions in the nearby town, and providing psychiatric assessments in the ER of a general hospital. Although Roffey Park was not a teaching hospital, I received regular psychotherapy training and supervision sessions by a visiting psychotherapist from St. George's Medical School in London. The consultant psychiatrist who took care of me was of Indian origin, a well-respected and experienced psychiatrist whom everybody addressed as "Raj" and whom I will call Dr. R. Although I arrived without any training in psychotherapy, he was pleased to have a young MD from Switzerland on his team. I guess he thought I had psychotherapy in my blood, coming from the country of Carl Jung and next door to Austria, the country of Sigmund Freud.

This must have led Raj to make a blunder very early in my training. He assigned to my psychotherapeutic care a forty-two-year-old, severely depressed woman, married, with two children, who had spent five weeks in hospital without getting better. The fundamental mistake he made was to stop her antidepressant medication in spite of the persisting typical symptoms of a severe depression.

I started to see her twice a week. She didn't get better. This is how in a later publication I described my experience:

> The first suicide in my second year of training was a blow that hit me totally unprepared. When I arrived at the hospital the patient's body had been taken away by the ambulance people and the police, and the head nurse asked me if I would inform her husband. My mind went blank. I was unable to do so. I was so shocked that I couldn't think clearly and didn't feel I could speak with family. I felt that I failed in my role as a doctor when I asked the head nurse to take over. The consultant psychiatrist, a normally soft and caring man, later that morning wanted to know what had happened, looked hurt and walked away. The case was never mentioned again. There was no team supervision, no case

review, no debriefing. I don't even know if the consultant ever spoke to the patient's husband. I reevaluated the case with my psychotherapy supervisor, but he did not follow up with me about the case. A few days after the suicide I developed a paranoid fear of her husband whom I had seen once with the patient. The thoughts that one day he would turn up and assault me or even kill me became more and more intrusive. I developed a fear of the dark: At night I got frightened to leave the house, especially when I had to go to the hospital when on duty. I never mentioned these fears to anyone, not even to my wife. I would have been ashamed to admit such irrational thoughts.

At that time there were several suicides per year in that small psychiatric hospital. In a way, what I learned as a young trainee in psychiatry was that suicide must be part of psychiatric routine and that, as in any accident/emergency unit, medical staff were simply expected to be strong enough to deal with this on their own. Maybe not surprisingly, two years later I started my first research project on suicide risk factors.[1]

Much later I read somewhere: There are two kinds of psychiatrists—those who have had a patient die by suicide and those who will. I wish I had read that at the time.

Dr. R. couldn't deal with the shocking outcome. He looked hurt, and he never talked about this case. I can only assume what he felt. And I was left alone with my feelings of guilt and failure.

Indeed, the suicide of a patient is a serious blow to the professional identity of any health professionals involved. Some young psychiatrists leave psychiatry altogether after such an experience, now being afraid of suicidal patients. The late Jerome A. Motto, a professor of psychiatry in San Francisco and a wonderful clinician, wrote one year after the suicide of Gil: "The most vulnerable therapists seem to be, understandably, those who are in the early period of their professional career."[2] Psychological care after the suicide of a patient is essential for the professionals involved. The suicide of a patient in a hospital can deeply affect an entire multidisciplinary team and requires team counseling sessions. Team members' emotional responses may include guilt, distress, anger, panic, sadness, self-doubt, and rumination but also fear of blame and fear of suicide of other patients. In the aftermath of all suicides, the people involved search for explanations and answers to the question whether they have missed something. Recommendations for mental health nurses include the "need to establish and promote a culture of openness in which suicide is anticipated as a possible outcome, even with excellent standards of care and wherein all staff are supported and encouraged to discuss and reflect on their concerns and fears during the aftermath of a client suicide."[3]

Despite the lack of support in the case of Gil, Dr. R. had a positive and long-lasting effect on my professional development. Early in my time at Roffey Park Hospital he said to me, "Konrad, you must never forget that every good health professional, besides his clinical work, always does some research. Don't take things for granted and always reflect about your work." This made me apply for a job in a research department in Oxford. Although the Oxford chair of psychiatry, Michael Gelder, told me that with my training I did not qualify for an academic career, he wanted to give this Swiss trainee the chance to get some experience and offered me a research job for a year. Tamara Kolakowska, the head of the research unit at the Littlemore Hospital, took me under her wing, especially with my first publication, about patterns of drug prescribing in two psychiatric hospitals, and with her great help—she rewrote most of my text—this manuscript made it into the respected *British Journal of Psychiatry*. I came to admire the clarity and succinctness of English scientific writing, but I still had a long way to go: When looking through the essays I wrote as preparation for the final exam of the Royal College of Psychiatrists, Prof. Gelder would more than once say: "Konrad, I know how difficult it must be for you with your background of German muddled thinking. . . ." Two years later I returned to my home country; I was a proud, UK-trained psychiatrist and member of the Royal College of Psychiatrists. Switzerland at that time did not yet have a structured curriculum for training in psychiatry.

The huge difference between the suicide rates in England and Switzerland was striking and unexpected. In 1981, the suicide rates were 14.5 per 100,000 in England and Wales and 23.8 per 100,000 in Switzerland. With my new diploma from the Royal College of Psychiatrists, I thought that the high suicide rate in Switzerland might be related to the poor knowledge of my colleagues in the assessment and treatment of suicidal patients. Maybe the suicide rate in Britain was lower simply because British doctors were better trained?

How do health professionals recognize patients at risk of suicide? The leading opinion in psychiatry was that over 90 percent of adults who die by suicide have a psychiatric disorder; that is, they are mentally ill. These findings had been corroborated by several researchers since the 1970s. The general message from these studies was that suicide is the consequence of a psychiatric disorder, mainly affective disorders, such as depression.

Suicide prevention therefore must focus on psychiatric pathology and the adequate treatment (usually medication) of mental disorders.

With my newly acquired knowledge I wanted to reach a wider audience in my home country. I wrote an article for Switzerland's main weekend magazine: "Suicide Doesn't Have to Be" ("Suizid muss nicht sein"). The article focused on the association between suicide and depression and the importance of recognizing and treating depressive disorders. Two weeks later, the magazine ran a prominent full-page letter to the editor by a Swiss writer and philosopher:

> How ridiculous to reduce suicide to a malfunction of the brain, to see desperation as a fault in the system, something that can be corrected with drugs. Suicide as an error, a mistake, an overreaction. Humans as machines, medicine as a repair business. How obscene. I hear from afar Jean Améry's laughter—he who knew so much about humans and their suffering. Never ever would I go to see Dr. Michel, and I would not recommend it to anyone else. He would suppress the symptoms of depression with antidepressants, and he would not help me to understand the meaning of my depression—which certainly is more than just an illness. . . . The key question is: What right do we have to decide about an individual's life (and death), to keep someone alive against his will, to coerce someone into continuing his or her life? For the person concerned suicide makes sense, and when life is not worth living any more, the doctor has to accept this and let the person go.[4]

Of course, this assault on my "royal" training did not occur without denting my young professional self-esteem. But it did not shake my belief in what I had written. After all, there was my

British diploma. I was convinced that the number of suicides could be reduced and lives saved with better recognition and treatment of depressive disorders.

Let me add that years later, with a wider horizon on the phenomenon of suicide, I would have loved to publicly discuss the issue with August E. Hohler, the author of the letter. I believe it could have been a constructive exchange of views.

My first study on suicide risk factors was triggered by a research paper published in 1975.[5] The authors of this study compared the personal characteristics of sixty-four individuals who had died by suicide with those of nonsuicidal patients. The main factors associated with suicide were symptoms of depression, male sex, living alone, self-neglect, and a history of suicide attempts. I decided to repeat the study in Switzerland. If the conclusion was correct, it meant that there were clear risk factors that could help clinicians detect suicide risk in their patients.

I wrote a draft of a study protocol. The study should demonstrate the importance of proper assessment of suicide risk factors. Ultimately, my hope was that suicide in Switzerland could be reduced by improving physicians' skills in assessing suicide risk. The plan was to compare the typical characteristics of people who had made suicide attempts with those of people who had actually died by suicide. In order to have the same sources of information, I interviewed the relatives of both groups, plus the medical doctors involved in the patients' treatment. I proudly handed the project description to the secretary of the head of the psychiatric department in Bern and asked for an appointment with him. I was given an appointment in six weeks. Six weeks! I had expected high praise for my initiative. After all, most of my colleagues had no interest in research. Finally, the professor asked me in and offered a seat. "Well, Dr. Michel, what can I do for you?" I told him about my research project and the document he had received. It was obvious that he had not looked at my project description. I didn't know then that later we would become dear friends.

The important thing was that I was allowed to go ahead with my study project in addition to my clinical work with outpatients. This was a time when permission from an ethical committee was a minimal formal matter. To get the names of suicide attempters I went to the emergency department once a week and, with the help of the nurses, looked through their patient records. The names of people who had died by suicide were provided by the Institute of Forensic Medicine of the University of Bern. No ethical committee objected to my calling the people and families concerned, asking them to meet me for an interview. Quite extraordinary, considering the confidentiality issues from today's point of view.

I interviewed relatives of one hundred people who had either attempted suicide (fifty) or died by suicide (fifty). For most interviews I was invited to the families' homes, often in the evenings. In addition, I interviewed all primary care physicians and psychiatrists who had been involved in the care of these people. After some ninety cases I found it increasingly hard to pick up the phone and call yet another family left bereft by the suicide of a son or spouse or parent. I heard a hundred and more stories and was stunned by how individual and different they were; some of them I still vividly remember forty years later. The study was a true immersion into complex relationships, traumatic experiences, vulnerabilities, and unmet emotional needs. Indeed, it was an experience that turned out to provide a major motivation that has always kept me going as a psychiatrist dedicated to clinical work: the richness and incredible uniqueness of peoples' stories.

However, talking to the physicians and the psychiatrists who had seen these patients, I heard another story. Up to 50 percent said that they had been surprised by their patients' suicide. Some 40 percent said they'd had no knowledge of the past suicide attempts of their patients. Suicide had not been addressed in consultation.

Insight gained: Patients don't talk—and doctors don't ask—about suicide.

Few researchers at that time paid attention to the problem of communication between the suicidal patient and the physician. One whom I have always admired for his publications is George E. Murphy, from St. Louis, Missouri, author of "The Physician's Responsibility for Suicide: Errors of Omission."[6] Murphy looked at the treatment of patients who had died by suicide while under medical care. In only two-fifths of the cases was a history of attempted suicide known to the responsible physician, "despite the information being available from other sources." Murphy had pointed out some years earlier that he had found not only that there was "substantial evidence of depressive illness in three quarters of the patients" but also that the diagnosis was rarely made and therefore depression rarely treated. In an earlier paper, "The Physician's Responsibility for Suicide: An Error of Commission," he had written that over half of those who died by overdose had received a drug prescription a week or less before their death or were given an unlimited prescription for a lethal amount of a hypnotic medicine.[7]

In my interviews with the physicians, I found that, indeed, depression had very often not been recognized—exactly what Murphy had written about. But my interviews revealed another problem. There was serious confusion with the diagnosis of depression. Depression at that time was classified as reactive, neurotic, or endogenous (i.e., mainly genetic). This meant that *symptoms of depression* were mixed with the *assumed causes of depression*. The diagnosis largely depended on the psychiatrist's personal interpretation. If a clinician didn't see a trigger for the onset of the affective disorder, the patient would be diagnosed with endogenous depression—in other words, it was considered to be of some unknown genetic and biological origin, and the indicated treatment was drugs and ECT.[8] In contrast, a depression with an apparent external trigger was usually diagnosed as reactive, which meant less severe, more psychological than biological in its origin, and with poor response to antidepressant medication. In my study, a primary care physician of a patient who had died by suicide told me: "Well, Dr. Michel, you see, this man had a reactive depression after he had lost his driving license after driving drunk. I gave him sleeping pills because of the transient problems with his mood." The family told me that for weeks the man had stayed in his room all day, had withdrawn from all his social contacts, had lost interest in his usual activities, had lost appetite, didn't sleep, couldn't concentrate, and complained that his memory had gone. In short: the typical symptoms of a severe depressive episode, today a clear indication for antidepressant medication—not just for sleeping tablets.

I found that others had recognized the problem with the diagnosis. One publication I have always admired was by Jan Fawcett in 1972. He wrote: "The presence of a 'reason' for depression does not constitute a reason for ignoring its presence."[9] As we have seen with the case example just described, there was a personal negative experience that had preceded the onset of depression. The doctor obviously decided that this was a case of a reactive depression, that is, transient and not serious. Prescribing sleeping pills was a logical conclusion. Fawcett's paper left a lasting impression on me. The consequence of the family doctor's interpretation of the depressive state was that the patient did not receive the necessary antidepressant treatment. The man died by suicide.

"Konrad, again you are mixing up the description of symptoms and their interpretation!" Prof. Gelder in Oxford taught me that in case formulations the *description* of a phenomenon (that is, the signs and symptoms) and the *interpretation* (the assumed cause) of the observed phenomenon were separate entities. The Bible for British-trained psychiatrists was Karl Jaspers's *General Psychopathology* (*Allgemeine Psychopathologie*). It was the heaviest volume on the official reading list for the diploma of the Royal College of Psychiatrists. Jaspers, a German psychiatrist and philosopher, argued in his 748-page work that in order to reliably and objectively diagnose psychological phenomena, the examiner had to rely on what the patient communicated and what the examiner observed—that is, on descriptive psychopathology, which is based on phenomenology, a concept described by the philosopher Edmund Husserl in 1901. Phenomenology aims to come to a value-free understanding of the patients' personal experiences by avoiding the interviewer's explanations of psychopathology.[10] "Empathy is often itself responsible for 'psychologizing.' There is a desire to 'understand' everything and all critical awareness of the limits of psychological understanding is lost. This happens whenever 'psychological understanding' is turned into 'causal explanations' under the misconception that in every case there is a meaningful determinant of experience to be found."[11]

The danger of explanatory theories may be demonstrated with the (mis)use of the diagnosis of schizophrenia in the USSR, which can be traced back to Swiss psychiatry.[12] Eugen Bleuler, the eminent psychiatrist and director of the Burghölzli psychiatric university hospital in Zürich from 1898 to 1927, first used the term *schizophrenia* in 1908, replacing Kraepelin's "dementia praecox." Bleuler described different forms and symptoms of schizophrenia and with these categories tried to further the understanding of this enigmatic mental disorder. The concept of a hidden or latent form of schizophrenia relied on the idea that in some cases the symptoms were discreet and not yet apparent. In the old USSR this diagnosis was used to lock away political dissenters in psychiatric hospitals, suffering from "slow progressive or sluggish schizophrenia," with "antisoviet tendencies" and "reform delusions," clearly a political, value-based concept. It is estimated that some two million dissenters were removed to psychiatric institutions.

Insight gained: In psychiatry, theories based on subjective interpretations can have serious consequences.

Fortunately, the classification of mental disorders underwent remarkable changes, mainly as a result of the American *Diagnostic and Statistical Manual of Mental Disorders* (*DSM*). While the early editions reflected a predominantly psychodynamic psychiatry, for instance, with the use of the term *neurosis*, the main achievement of the *DSM* classification was to free the diagnostic criteria from causal assumptions—that is, to keep the diagnostic criteria on a descriptive (phenomenological) level. Over the years, the *DSM* classification system has developed and is currently in a fifth edition (2013). Neurotic disorders, which were associated with the psychoanalytical concept of neurosis by Sigmund Freud and Carl Jung, now became anxiety disorders; depressive disorders are now categorized as mild, moderate, or severe (recurrent or persistent); schizophrenia has lost its subtypes; and so on.

But let me return to the early 1980s, where in my interviews with primary care physicians and psychiatrists I had stumbled over the problems with the diagnosis of depression. As an enthusiastic young doctor I wanted to have an effect on medical practice in Switzerland, so I summarized the results of my study about the one hundred cases of suicide and attempted suicide for the *Swiss Medical Weekly*: "Could the Medical Profession Do More?"[13] I referred to my main findings: the problem of unrecognized depression, the lack of knowledge of suicide risk factors, and deficits in the communication between doctors and suicidal patients. The main publication followed later in the *British Journal of Psychiatry*: "Suicide Risk Factors: A Comparison of Suicide Attempters with Suicide Completers" (terms not recommended any more these days!).[14] I reported that suicide attempters who had used medically serious methods had more depressive symptoms, had used more life-threatening suicide methods, and more often had a history of suicide attempts. These suicide risk factors have been confirmed by many other studies. I argued that the knowledge of these factors would allow doctors to recognize patients with a high suicide risk.

Prof. Battegay, from Basel, a gray eminence of Swiss psychiatry and past editor of the professional journal *Crisis*, one of the early suicide-focused journals, attended one of my very early presentations. He asked me to write a paper for *Crisis* from my study with one hundred patients. Here is what I wrote in the discussion section:

Suicide research has provided us with a fairly good picture of the typical characteristics of the average person who puts an end to his or her life. The typical patient is male, over 40, likely to be suffering from depressive illness, and he may have an alcohol problem or abusing drugs. He probably has a history of prior suicide attempts, and he is likely to have given warning about his suicidal thoughts to relatives and friends. He is under medical care, although his psychiatric problems are not necessarily a focus of treatment.[15]

Clearly, my first studies were mainstream psychiatry, dealing with suicide risk factors and the relationship of mental disorders with suicide.

SUMMARY

In this first chapter we have seen that the suicide of a patient had a long-lasting effect on my professional career. I am by far not the only one with this experience. I have met colleagues who gave up psychiatry altogether after the suicide of one of their patients. For me, the difference in suicide rates between the United Kingdom and Switzerland was a challenge to look into the role of medical practitioners in preventing suicide. Assessing suicide risk based on risk factors has traditionally determined the medical approach toward suicidal patients in emergency departments and psychiatric hospitals. This approach guided me in my study of one hundred cases of suicide and attempted suicide. However, it also opened my eyes to the problems with the psychiatric diagnosis and the fact that suicide was often not addressed in medical consultation.

2

FIRST LESSONS IN REDUCING SUICIDE

In 1989, Wolfgang Rutz and colleagues published the results of the Gotland Study.[1] In this elegant study, primary care physicians on the Swedish island of Gotland in the Baltic Sea, with 56,000 inhabitants, were offered an educational program aimed to improve the detection and diagnosis of depression and its adequate treatment. The effect on suicidal behavior was spectacular: Suicides decreased from a calculated rate of 25 cases/100,000 in 1983 to 14.3 in 1984 and to 7.1 in 1985. In the same period, numbers of prescriptions for antidepressants went up, prescriptions for benzodiazepines dropped, and fewer patients were admitted to inpatient care. These were truly exciting results. It meant that improving the primary care providers' clinical knowledge and skills could save lives.

I decided to train medical practitioners in Switzerland. First, I wrote a research plan on how I could improve physicians' knowledge and attitudes related to suicide and depression. Research here means evaluating the effect of what you do. I developed a multiple-choice questionnaire, which I sent to medical practitioners. Knowledge questions included items about suicide in Switzerland, suicide risk factors, and symptoms of depression. More interesting were the items related to attitudes toward the physicians' perceived role in suicide prevention. The physicians had to indicate how much they agreed/disagreed with the following ideas:

- Each individual has the right to kill themself
- It is questionable to interfere with the decision of a person to kill themself
- Suicide is always the expression of a mental disorder
- Medical doctors can prevent suicide
- I think I would recognize a suicide risk in my patients
- I feel competent in the management of suicidal patients

An impressive 60 percent of recipients returned the questionnaires. The responders were then divided into two groups. One group received an educative handout; the other was offered a seminar-style single session of training, in addition to the handout. The handout had the title "Suicide: A Challenge for the Physician." I drove around the canton, mostly in the evenings, meeting with small groups of practitioners at someone's home. The three-hour meetings involved teaching and case discussions, followed by food and wine. What aided my rapport with the physicians I met with was that I had always respected and admired medical practitioners working as dedicated family doctors in rural areas like the Emmental (the origin of the famous cheese). In medical school, my dream had been to establish myself as a primary care physician in an area like this. How wonderful, I thought, it would be to spend a professional life as a caring family doctor for a rural population. Yet, after two years of internship in medicine I found myself drawn to psychiatry. I was fascinated by the human, philosophical, and intellectual challenges of psychiatry and the one-to-one therapeutic involvement with patients—something I had missed in medicine.

In my many visits to the doctors' peer groups I heard many stories that illustrated the difficulties physicians had in dealing with suicidality. Here is one of the stories:

> A primary care physician from a rural practice told the group the following story. On a busy Saturday practice morning a 45-year-old teacher whom he has not seen for two years presents with a strained left ankle. The patient reports that it happened a couple of days ago when he went for a walk in the forest. In the examination the doctor cannot find anything special, and he discharges the patient with an ointment and an elastic bandage. Two hours later the patient's wife calls and asks if her husband is still in the surgery as he had not returned yet. An hour later she calls again. Her husband was found dead in the forest. He had shot himself through the head.

Although people who have suicidal thoughts and plans very often are under medical care, they rarely talk about it. Researchers from Finland found that at the last visit before suicide, the issue of suicide had been discussed in only 22 percent of the cases (39 percent in psychiatric outpatients, 11 percent in general practice).[2] Some studies found an increase in the frequency of visits to medical practitioners before suicide and attempted suicide, suggesting that the reasons for the visits to care providers are related to the patients' mental problems and suicidality—but even then, suicide was too often not addressed in consultation. In fact, as in the case example just presented, it is not at all rare that patients visit their doctor a few hours before their death. In the Finnish study, 18 percent of those who had contacted a physician had done so on the day of their suicide, yet even then the issue of suicide was rarely discussed.

After I had finished my teaching visits I sent a follow-up questionnaire to every physician who had attended my seminars and to those who had only received the written material. Clear differences between the two groups emerged. In comparison to the handout-only group, the seminar group had developed a more active, less defeatist attitude in the clinical prevention of suicide. They had better knowledge and felt more confident in recognizing suicide risk. My conclusion was that written material alone did not have any effect at all on knowledge and attitudes.

Insight gained: Effective teaching requires personal interaction.

To publish the results, I needed help with the statistics. This is where the collaboration with my colleague Ladislav, a gifted statistician and an ingeniously creative psychologist, began. I chose a somewhat bold title for our manuscript: "Suicide Prevention: Spreading the Gospel to General Practitioners." The paper was published in the *British Journal of Psychiatry*.[3]

Today, some excellent programs for primary care providers ("gatekeepers") are available, targeting knowledge, attitudes, and interviewing skills.[4] For instance, the QPR is a one-to-two-hour program designed to teach gatekeepers emergency response skills to identify and intervene with someone at risk for suicide and how to get help for an individual at risk for suicide.[5] However, I am sure the problem remains that, generally, only a small proportion of general practitioners attend workshops and trainings for dealing with suicidal patients.

Improving the knowledge and attitudes of health professionals in dealing with suicidal patients was the key issue in a prevention project we launched in 1990 in Switzerland. Conrad Frey, a friend and member of the executive committee of the Swiss Medical Association, joined the project group and secured the support of the Federal Office of Health. A working group with experts from adult and adolescent psychiatry, preventive and primary care medicine, and a lived experience organization was set up. A basic outline of the "philosophy," the objectives, the target population, and the didactic materials was formulated. To emphasize that suicide risk is an acute and personal condition, the campaign was called "Crisis and Suicide" ("Krise und Suizid"). As by now I had learned that written material alone does not change anything, we decided to use an interactive teaching method similar to the Gotland study. We trained trainers, always in pairs of primary care physicians and psychiatrists. We developed a brochure for the participants and a manual for trainers.[6] A main objective was to facilitate the communication between suicidal patient and the physician. For this we produced videos with role-plays of general practitioners interviewing patients, as well as leaflets for the doctor's waiting room, encouraging people to talk to their doctor about suicidal thoughts. Pocket cards for the physicians, listing the main depressive symptoms, with model sentences for exploring hidden suicide risk were distributed, and a series of suicide-related articles for the *Swiss Medical Journal* were commissioned.

Introducing the term *crisis* into the vocabulary of suicide prevention meant that we chose to go beyond the medical model. Crisis is not a psychiatric diagnosis. The term *crisis* stems from the Greek word "*krisis*" and means "decision or turning point." The concept of crisis and crisis intervention is associated with the name of Gerald Caplan and his *Principles of Preventive Psychiatry*.[7] To our surprise, in the workshops the medical practitioners said that the crisis concept was new to them; they had not come across it in their training. In our campaign, "crisis" was defined as "a time-limited disturbance of the psychological equilibrium of the individual." It was important for us to stress the individual, subjective, and transient nature of this experience. A crisis has a triggering event—an initial, identifiable occurrence in the life of the individual. Triggers can vary drastically in scale, from large-scale natural disasters and war to situations that may appear less dramatic (e.g., incidences of bullying in a school, marriage, transition from college to a job). The one thing that is important about an event that precipitates a crisis is that it is perceived as a threat to the emotional stability of the individual. The crisis state is different from other types of life stress because the outcome may be maladaptive; that is, a negative development may ensue. The person in crisis is more vulnerable and less resilient. In our campaign, psychiatric diagnoses and other risk factors were kept on board, but—and this was new—they were now labeled as risk factors, allowing room for the patient's subjective experience.

Besides the crisis concept, we emphasized another basic issue: "The recognition of a suicide risk is a matter of communication. Doctors and patients must learn to talk about these important matters." Little did I know then how much this sentence would be the program for my future professional work.

In our workshops, physicians learned that admitting a patient to inpatient care does not mean that the patient will be safe, because suicides happen in hospitals quite frequently.[8] Instead, we aimed to strengthen the role of primary care physicians in suicide prevention (remember: in the Gotland study, fewer patients were admitted to inpatient care). A key message was:

"A good therapeutic relationship is the best anti-suicidal means. The suicidal person above all needs someone who tries to understand and who can accept and bear the despair. . . . To tell the person what he or she should do is usually not helpful."

The campaign was well received, but it soon became obvious that suicide was not a priority issue for medical practitioners. Improving the knowledge about treatments for headaches or hypertension was more important. Out of some six hundred invitations to primary care physicians, fewer than ten would participate in one of the—free of charge!—two half-day workshops. Those who had participated as co-trainers reported that once they had started asking their patients much more frequently—and with more ease—about suicidal thoughts or behavior, they were utterly surprised how often patients revealed suicidal thoughts or past suicidal behavior. Originally, as part of the campaign, I had planned to replicate the Gotland study, taking the rural Appenzell canton as an equivalent of the Baltic Sea island. The idea had to be abandoned when we found out that a large proportion of that population got their medical care in the neighboring cantons—and after we had learned how difficult it was to get general practitioners to attend our wonderful workshops. My enthusiasm got slightly deflated. I concluded that, although training medical practitioners in recognizing and treating suicidality in their patients remained an important element in suicide prevention, it was not the way to effectively reduce suicidal behavior in the society.

Investigating the role of media reporting on suicide was an important module of the Krise und Suizid prevention campaign. Today, it is widely established that media reporting on suicide can increase suicidal behaviors. What matters is the amount and prominence of coverage, with repeated coverage and "high-impact" stories being most strongly associated with suicide and attempted suicide. The so-called Werther effect is well known: In 1774, Goethe published *Die Leiden des jungen Werther* (*The Sorrows of Young Werther*), in which the protagonist shoots himself because of his unrequited love for Lotte, who is engaged to Albert. The novel caused a wave of suicides across Europe following its release—many of those who died were dressed in a similar fashion to Werther, adopted his method, or were found with a copy of the book.

This component of the Krise und Suizid campaign turned out to be a rather big thing. First, with the help of a media agency called Argus, every article relating to suicide in one way or other from daily newspapers and magazines in the German, French, and Italian languages was collected over a period of eight months. There was no digital access to print media in 1992, so we ended up with heaps of paper clippings. Each article was rated using a sophisticated, fifty-item questionnaire we developed based on recommendations from a U.S. national workshop on the reporting of suicide in 1991.[9] We used these recommendations to create an "Imitation Risk Score." The underlying assumption was that the risk of contagion would be higher the more attention an article received (the article being on the front page, the word "suicide" appearing in the heading, photos of the person or the scene, the victim being glorified, etc.).

This is what we found: Nearly 10 percent of the suicide articles were placed on the front page, 50 percent had sensational headlines, and 40 percent included a picture of the suicide victim or the place of death. Driving these figures was the Swiss tabloid *Blick*, the print magazine with the highest national circulation figure (figure 2.1). Several times a year it would place

FIGURE 2.1. "Love pangs? Girl (17) jumps down onto highway."
Source: *Blick* (October 1992).

an article about a teenage suicide on the front page, usually with sensational headlines ("Bad Marks at School—12-Year-Old Kills Himself").

Today I remain proud of my idea to arrange an official meeting with the chief editor of *Blick*," Mr. Luchsinger, with a clear strategy in mind. We told him that we were going to present the results of our analysis in a big national press conference launched by the Swiss Medical Association and that his newspaper would be *the* main culprit. But, we added, there would be a way to change things: He could help our cause by publicly agreeing to follow the guidelines for suicide reporting. It was a tough, two-hour discussion in the chief editor's office. He said, "It's my job to personally decide what goes on the front page. 'Dog Bites Man' is not interesting, but 'Man Bites Dog' is." I pointed out that this argument didn't hold, with nearly two hundred young people in our country dying by suicide each year. Finally, he gave in and agreed to sign the document we had prepared. "I shall in the future adhere to the guidelines for reporting on suicide." We left the publisher's building in Zürich and headed straight to the nearest pub. We had earned a drink.

With the press conference, guidelines for responsible suicide reporting were issued in three languages, distributed to journalists attending the conference and sent to every newspaper's editors. Most print media reported on the results of the study and the risk of imitation ("Medical Doctors Issue a Warning About the Danger of Sensational Reports on Suicide," etc.). Swiss TV included an interview with Mr. Luchsinger on the main news program: "Well, I wasn't really convinced by the arguments of these young doctors, but I finally agreed that in the future we are going to follow these guidelines."

We assessed the effect of the media campaign in a second analysis of Swiss print media eight months after the press conference. The percentage of headlines on the front page was clearly lower, and they were rated less often as sensational or glorifying. Fewer articles had pictures, and generally the reports were less sensational. Altogether, in the second analysis, the Imitation Risk Score was significantly lower for the seventy-four main newspapers, and in particular for *Blick*. Rather amazingly, the scientific article published in *Crisis*, "An Exercise in Improving Suicide Reporting in Print Media," remains one of the very few reports showing a significant effect of media guidelines on suicide reporting.[10] I like to believe that the intervention with the chief editor made the difference. However, I must admit that I got the idea from Gernot Sonneck, a dedicated psychiatrist from Vienna who had told me earlier that personal contact with newspaper editors was absolutely crucial. In 1994 Sonneck had reported the results of a landmark study in which he showed that changes in how the media reported on Vienna Metro suicides resulted in an impressive 75 percent drop of suicides, which was sustained for more than five years.[11] He told me that he would call editors immediately whenever he came across a "forbidden" sensational article. Maybe the lack of personal contact with newspaper editors is the main reason why most studies investigating the effect of guidelines for media professionals have not found an effect.

Insight gained: Successful implementation of guidelines for suicide reporting requires personal contacts with media professionals.

Today, responsible media reporting is established as one of the important elements of national suicide prevention strategies. Various guidelines for media reporting on suicide have been published worldwide (see https://reportingonsuicide.org/).

The huge differences in suicide rates between the European countries lingered on my mind. How on earth could these baffling differences be explained? Wouldn't research allow us to better understand suicidal behavior and eventually also help us to become more effective in reducing suicide?

I wrote a letter to the Swiss Federal Office of Health, inquiring if they were interested in a comparative international study project aimed at looking into the factors that contributed to the differences in suicide rates. I got a call two days later. The caller remembered that some months ago they had received an invitation for representatives from Switzerland to join a multinational project initiated by the WHO Regional Office for Europe, with the goal of monitoring suicidal behavior in Europe. They had not answered to the letter (typical for Switzerland); nobody was interested in an international study project. I was told that the first official meeting was to take place in two weeks, Edinburgh. If I wanted to attend, they would see if it was possible to cover the expenses.

In Edinburgh I found that most participants were members of an international working group whose task was to develop detailed plans for a coordinated multicenter project. Although I was a novice in international circles, I was warmly welcomed, and in the evenings, I frequented the local pubs with the others. I learned that behind the initiative for a collaborative multinational research was the WHO target no. 12, which the member states—including Switzerland—had agreed on in 1984: "By the year 2000, the current rising trends in suicides and attempted suicides in the Region should be reversed." I got to know people like Ad Kerkhof from the Netherlands, Stephen Platt from Scotland, Unni Bille-Brahe from Denmark, Armin Schmidtke from Germany, Diego De Leo from Italy, Christian Haring from Austria, Paolo Crepet from Italy, Jouko Lönnqvist from Finland, Danuta Wasserman from Sweden, Tore Bjerke from Norway, and, over the years, many more who joined the multicenter study group. At the end of the Edinburgh meeting the WHO/EURO Multicenter Study Group on Suicidal Behavior was formed, and for many years it would keep me involved in a very active international research cooperation, with regular meetings all over Europe. It felt like a large patchwork family, with many wonderful friendships—as well as some major conflicts. The first multicenter project to be developed was the "Monitoring Study," a first attempt to collect comparable data on suicidal behavior across Europe.[12] The second multicenter project was the "Repetition-Prediction Study," aimed at identifying the social and personal characteristics predictive of repeated suicidal behavior.[13]

Bern, with a metropolitan-area population of 320,000, became the site of the Swiss collaborating research center. In the first twelve months we recorded 243 suicide attempts carried out by women and 153 by men. The combined statistics came up with 177 suicide attempts per 100,000 people, putting Switzerland relatively close to the overall European rate of 129 suicide attempts per 100,000 inhabitants. Certain points from the first reports of the results are worth mentioning. The yearly rates of attempted suicide varied widely between the fifteen centers: Helsinki (Finland) had a rate of 414 suicides per 100,000 inhabitants, while Leiden (Netherlands) had just 61 per 100,000. Most methods used for attempted suicide were—fortunately—not immediately lethal methods, usually overdosing or cutting, which means there is time to intervene. High percentages of prior suicide attempts were found and a repetition rate of 15 percent in the year after an attempt.

Collaborating centers could use the growing data pool to analyze certain factors related to suicidal behavior. We chose to study the drugs used for overdosing. England (Oxford) had a rate of painkiller overdose many times higher than other centers. In a meeting, Keith Hawton from Oxford claimed that my results were wrong. But they weren't. The United Kingdom at that time had a huge problem with paracetamol overdoses (which may lead to fatal liver failure). This finding eventually led to changes in the availability of painkillers, which so far had been available without medical prescription in shops, even in gas stations, in bottles of one hundred tablets. In 1991, the second stage of the WHO/EURO multicenter study started: the Repetition Prediction study. Patients who had attempted suicide were interviewed using the European Parasuicide Interview Schedule, called EPSIS. In Bern, most of these interviews were carried out by trained psychology students and took up to four hours (!) to complete. One year later, patients were contacted again, and forty-eight out of sixty-six were re-interviewed.

As a clinician, I was not happy with this project. The EPSIS did not leave any room for the suicidal person's individual account of their suicidal crisis. My psychotherapy training had taught me how crucial it is to listen to patients and to understand their personal logic. Unlike most of my colleagues in the multicenter study, I had received a fairly robust training in psychotherapy, and I was working in a psychiatric institution whose medical director was a highly respected psychotherapist. After all, my motive to become a psychiatrist and psychotherapist had been to get involved with patients, with people.

Insight gained: Collecting data on suicidal behavior is important but doesn't help us understand the patients' reasons for suicide.

But what could be the inner logic of a person considering suicide? A paper published by Bancroft and colleagues was an eye-opener, in that, once more, it showed the fallacy of interpreting the suicidal patients' behaviors.[14] The authors had interviewed 128 "suicidal subjects" immediately after their recovery from an overdose. The patients were given a list of possible reasons for taking overdoses and asked to choose any that applied to them. The statement chosen most often (52 percent) was "obtaining relief from a terrible state of mind," and 44 percent indicated that they had wanted to die. The authors then gave the same list of statements to the clinical psychiatrists and compared the answers. The differences were striking: "The situation was so unbearable" was chosen by 56 percent of the patients and 0 percent of the health professionals; "Lost control and don't know why" was 27 percent and 0 percent, respectively. "To make people understand how desperate you felt" was 20 percent versus 71 percent; "To influence someone to change their mind" was 7 percent versus 54 percent. The interpretation was pretty obvious: Patients primarily indicated *intrapersonal* reasons, while psychiatrists attributed *interpersonal* or *manipulative* reasons to the suicidal behavior. In the discussion of the results, the authors speculated about the reasons of these differences, and wrote: "It is noteworthy that the two least socially acceptable reasons, 'to influence someone' and 'to frighten or get your own back on someone' were seldom chosen (by patients). It is certainly possible that the reasons chosen have little relevance to the determinants of the act . . . but were chosen 'post hoc' (after the event), to enhance the social acceptability of the behavior." Obviously, the psychiatrists' explanations did not match the patients' explanations. With my training in psychotherapy, I didn't feel that it was helpful to assume that patients were liars—and, besides, why should they lie? I saw the patients' answers as the true expressions of their inner experiences in the suicidal crisis. Suicidal behavior had something to do with an acute and unbearable state of mind ("The situation was so unbearable") and a loss of control ("Lost control and don't know why").

It was truly surprising how few publications had dealt with the patients' own explanations of their suicidal behavior. Trying to understand what went on in the suicidal person's mind meant that in the Repetition Prediction Project I wanted to go beyond asking questions, considering the insight I had gained from my training in the Present State Examination (see chapter 1). I decided to add our own open-ended questions to the EPSIS interview, and we later published the results in a paper titled "Understanding Deliberate Self-Harm: The Patients' Views."[15] One of the questions was: "What made you harm yourself?" One-third mentioned relationship problems, 20 percent mentioned an "emotional crisis," 15 percent a mental disorder, and 8 percent "emptiness, no sense in life." I was struck by the low percentage of psychiatric illness in these answers, which was in complete contrast to the official medical and psychiatric opinion. I included the items from the Bancroft et al. paper, and we asked patients to recall how they had felt immediately before the event, that is, to indicate on a three-point scale how much they agreed with the statements presented to them. The results were interesting. The items with the highest scores were: (1) "The situation was so unbearable that I could not think of any other alternative"; (2) "my thoughts were so unbearable, I could not stand them any longer"; and (3) "I wanted to die." This meant that the overall picture therefore was similar to the Bancroft study: *Intrapersonal* motives were significantly more often chosen as of "major influence," while *interpersonal* items (e.g., "I wanted others to know how desperate I felt") were chosen by less than a third of patients.

But there was more to it than this. We asked patients—one year after their suicide attempt—about seeking help. The question was: "Who could have taken some action to help you?" To our surprise, 52 percent said "nobody." Twenty-one percent mentioned relatives or friends, and to our even greater surprise only 10 percent mentioned a medical doctor or other health professional. Fifty percent said that they could not have accepted help. But at the same time, 82 percent indicated that the suicidal crisis had had positive consequences for them, for instance, "Life is worth living" or "new start in life." In the related publication we wrote:

> Bancroft et al. discussed the possibility that the attempters' choice of motives might express a wish to gain social acceptability for their suicidal behavior. Another explanation is that the mental state immediately before the initiation of (para)suicidal behavior is indeed characterized by an acute state of emotional perturbation which the individual experiences as unbearable, and which has to be distinguished from the underlying—and often long-lasting—problems a person has to face. . . . Thus, the suicidal act is not—at least not primarily—an act aiming at

manipulating others, but a means of finding relief form an unbearable mental state.

It was clear that we had to put the patients' own personal experience first, if we wanted to learn more about the patients' emotional perturbations. Traditionally, in medicine the patient's views are "subjective" and are frowned upon, in contrast to the physician's "objective" assessment. But from my psychotherapy training I knew that the only way to engage a person in a meaningful therapy was to start from the patient's very own and personal inner experience. We were not the first ones to listen to suicidal patients and put their experience first. Here, I must make a reference to Edwin Shneidman, an American clinical psychologist. Together with Robert Litman and Norman Farberow in 1958 he had established the Los Angeles Suicide Prevention Center. His journey into the suicidal mind started with the study of several hundred suicide notes he had discovered in the vault of the Los Angeles Medical Examiner. It was Shneidman who as early as 1985 stressed the importance of emotional pain and of suicide as a solution for this unbearable condition.[16] Shneidman became the icon of American suicidology. His descriptions of psychological pain were and still today are masterly.

I realized that I had to work on my understanding of the suicidal person's "subjective" inner experience. What did "unbearable mental state" mean and what, then, did suicidal individuals need to stay alive? In 1995 I attended a meeting of Equilibrium, a Swiss self-help organization for people suffering from depression and their family members. During the lunch break, the organizers asked me to chair a session of a group with people who had a history of attempted suicide. There were some thirty-five people in the room, mostly women. I let them first talk about their suicidal crises. One of them, an attractive woman in her mid-thirties, took off her scarf and showed me a large scar on the side of her neck—she had tried to behead herself with a chainsaw. Such were the people in this group. I asked them: "Can you tell me who could have done something to stop you from doing something like this to yourself?" Their answers were unanimous: "We would have needed a person who would listen to us without trying to talk us out of it."

Extraordinary.

Obviously, not many clinicians could bear to listen to acutely suicidal people without starting to talk them out of it or thinking about when and how to call an ambulance or the police to admit the patient. But:

> It seems to us that communication problems between health care professionals and suicidal people seeking help are common, and that this may be a reason for some difficulties in the treatment of suicidal patients. When patients are not sure that they will be understood and accepted they are likely to keep suicidal thoughts to themselves. An obvious prerequisite for establishing a trustful working relationship is that patient and helper have a mutual understanding of the reasons for suicidal thoughts or self-damaging behavior.[17]

Psychotherapy research has established that effective therapy requires a therapeutic relationship based on a joint understanding of the problem. It was obvious: We needed a better conceptual frame to improve the understanding of the suicidal mind in order to overcome the problem of communication.

SUMMARY

In this chapter we have looked at the role of primary care physicians in reducing suicide. The effect of the Gotland study was impressive but never replicated. However, it was a trigger to teach medical practitioners in Switzerland. A major problem is that patients very often do not spontaneously mention suicidal plans—and medical professionals do not ask or do not know how to ask patients about their dark thoughts. In the seminars for practitioners we introduced the term *crisis*, thus attempting to lower the barriers for professionals and patients to talk about suicide risk. Teaching medical practitioners how to address suicidal thoughts and plans in their patients continues to be an important element in suicide prevention. In our early prevention campaign we included a study on suicide reporting in the print media. There, I learned that for interventions to be successful, personal contact with newspaper editors is crucial. The next important step for me was to join an international multicenter research project. However, as a clinical psychiatrist, I learned that collecting anonymous data from hundreds of patients does not really get us any closer to understanding the suicidal person.

3

EMOTIONAL STRESS AFFECTS BRAIN FUNCTION

There is a relatively easy method to record emotional arousal, familiar to most people from the lie detector used in crime stories, called electrodermal activity (EDA). EDA measures how easily electricity can flow between two electrodes placed on the person's palm. The electrical *conductance* between these two points depends on how much we are sweating. This is a physiological activity of the autonomous (sympathetic) nervous system, regulated by the brain, and reflects the emotional state of the person. EDA is extremely sensitive: It responds to minimal emotional arousal within 1 to 1.5 seconds.

One day, Ladislav, the creative colleague of mine, came up with the idea to use EDA to measure emotional stress related to our patients' suicidal crises. We fixed two electrodes to the palms of the patients we were interviewing after a suicide attempt. We also fixed electrodes to the interviewers' palms to see the interviewers' emotional reactions to the suicide stories. I was familiar with neurobiological studies demonstrating the existence of so-called mirror neurons in the human cortex, which fire in response to the observed actions of another person. They are thought to be the neural basis of the human capacity for emotions such as empathy.[1]

All sessions were video-recorded. We printed the EDA curves and pinned them on the wall together with the time-stamped interview transcripts. This allowed us to closely synchronize the spoken words with specific peaks in the EDA curves. This was a most fascinating piece of research. It felt like directly looking into the abyss of the emotional wounds that the person's life experiences had caused. Sometimes it made us feel like voyeurs. The EDA peaked when people talked about past traumatic experiences, for instance recalling incidents of domestic violence or separation. The EDA also peaked when they talked about the painful experience leading up to the suicide action. Of course, the EDA curve did not give us any information on the emotional *content* of the peaks. That part came from the transcript. It was striking that usually the emotional content related to the peaks was similar for both past and current experiences, such as traumatic separation or rejection. So, with our printed curves and interview transcripts on the wall, we visualized in a not very sophisticated way the connection between past emotional and recent suicide-related experiences.

Here is a sequence of the electrodermal activity of Ms. M., aged thirty-eight, a patient we shall meet again in chapter 7. She had made a serious suicide attempt by jumping from the third floor of a building. Her suicidal crisis started when her boyfriend said that he needed some distance from their relationship. Ms. M. was in a state of emotional arousal for days, with suicidal thoughts on and off. The last straw, however, came one morning when she accompanied her eleven-year-old son on the way to school; she drove, while he rode his bicycle. He unexpectedly took a shortcut, and she lost sight of him. Here is the EDA curve that covers her description of the "parting episode." The curves before and after this episode show the normal fluctuations of skin conductance (figure 3.1). For the middle section,

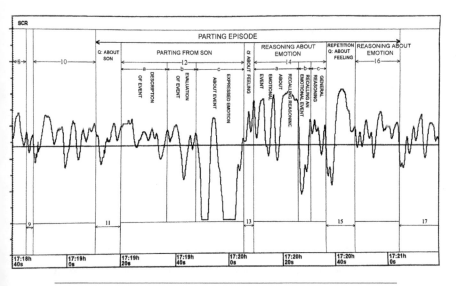

FIGURE 3.1. Electrodermal activity during the interview with Ms. M.

we see much bigger amplitudes. (Note: We do not distinguish between down or up peaks.)

Here is the story from the transcript of the narrative interview, related to sections in the EDA curve:

12 A: Ms. M. describes the situation of accompanying her son who then took another way,

12 B: she thought she wouldn't see him again, and hadn't even said goodbye to him,

12 C: she desperately wanted to see her son. Ms. M starts crying in the interview.

Then the interviewer asks her about her feelings.

14: Ms. M. starts reflecting about her emotions, recalling that for her it had been clear that she wouldn't survive the day, and *her children would be without a mother.*

16: things calm down again. She talks about going to her sister, but realizes that at that time she was not in control of her actions.

Clearly, the emotional issue is about separation. Later in the interview she talks about her lifelong fear of losing people she loves. In chapter 9 we will hear more about the story behind this life-endangering suicide action and see how it connects to an early traumatic experience.

Thinking of "mirror neuron empathy," together with the patient's EDA curve, we pinned the therapist's curve on the wall. In some cases there was a parallel arousal in both curves, as we had expected. My personal curve usually came out quite flat—which was quite embarrassing. My colleagues tried to reassure me. After all, they argued, a good therapist stays calm even with heavy stories of hurt and self-harm . . .

Although a stress response itself is unspecific, triggers, as we have seen, are specific. This leads us directly to *the concept of vulnerability and sensitivity toward specific triggers*—in the case of Ms. M. the fear of losing a close person.

Insight gained: Suicide trigger and personal vulnerability go together.

The interviews with EDA recordings were part of the early days of our search for neurophysiological correlates of suicidal behavior. We never managed to publish the results, as journal reviewers criticized our study method for not being scientific enough.

Remembering that life experiences are stored in our synapses, it was obvious that early stress-related memories—as a form of biographical vulnerability—and the acute suicidal mental state were psychological *as well as* biological phenomena.

From our interviews with suicidal patients we had learned that suicide attempts were usually carried out in an acute out-of-the-ordinary state of mind with a clear onset. Here, I need to introduce the concept of the *suicidal mode*. Aaron Beck, the American icon of cognitive therapy, had first used the term *mode* to describe how in depression thoughts, emotions, and behavior all change in a typical way, like a package, in synchrony. Later, David Rudd applied the concept to describe the "suicidal mode" (see box 3.1, "The Suicidal Mode and Dissociation").[2] Typical for the suicidal mode are sudden changes in thoughts, emotions, physiological reactions, and behavior. The presence of a set of explicit suicidal plans as the behavioral part of the suicidal mode presents a "readiness to respond" to events that trigger negative attributions about oneself (such as being worthless and the future hopeless). Because an act of self-harm (for example, an intentional overdose) often reduces the suicidal urge, self-destructive behavior is easily established as a response repertoire: *Suicide is a solution to an unbearable state of mind.* When triggered, there is an immediate "*switch-on*" of the mode, with the threshold becoming lower with each new activation. The concept of the suicidal mode corresponded well with how patients in the narrative interviews described the sudden mental changes they experienced in their suicidal crisis. We often heard people describe a "switch-on" phenomenon. I started to dig into papers on the technique of functional brain imaging. Would it be possible to see the switch from the "normal mode" to the acute suicidal mode in brain imaging?

I had come across studies that had used functional brain imaging for people diagnosed with post-traumatic stress disorder (PTSD). In a fascinating study, researchers had looked into the long-term effects of childhood sexual abuse on brain function in adult life.[3] They used a technique called positron emission tomography (PET) imaging in twenty-two women, scanning their brain activity while these women were listening to scripts of their personal experiences of abuse in childhood. PET imaging measures changes in blood flow in the brain areas. The researchers compared neutral scripts with traumatic scripts. The results showed that listening to scripts of childhood abuse was associated with a significant reduction of blood flow in a number of areas in the prefrontal cortex (PFC). In 2001, another research team used fMRI (functional magnetic resonance imaging),[4] a technique that measures neural activity based on blood oxygenation levels (representing metabolic demand). Here again, the researchers used script-driven recall of past traumatic events, comparing a group of study participants diagnosed with PTSD with a group of control participants. The PTSD group showed reduced levels of brain activation in the prefrontal cortex and some related brain areas.

The prefrontal cortex, the PFC, is the seat of problem solving, decision making, planning, and the organizing and monitoring of goal-directed behavior. In particular, the lateral frontal regions are closely associated with functions such as goal-directed planning, monitoring, inhibiting, selecting behavior, etc. The ability to reflect on a current emotionally stressful situation in the context of one's own autobiographical memory is largely dependent upon full operations of the PFC and its connections. The so-called ventromedial regions are involved in the regulation of emotional and behavioral responses. The PFC, as the highest cerebral region in guiding complex behavior, has connections to the emotional brain, that is, the limbic system and the amygdala. During uncontrollable stress, the activation in the amygdala, the emotional brain's alarm center, is turned up, while at the same time the function of the PFC is turned down (figure 3.2). This means that the problem-solving capacities of the PFC are severely impaired. Yet, the ability to put an acute emotional crisis into perspective and maintain long-term identity goals is crucial for coping with high levels of stress and mental pain.

A colleague of mine, Thomas Reisch, was highly motivated to go ahead with a functional MRI study with patients who had attempted suicide. Our plan was to use fMRI for investigating brain activation related to the suicidal mode, that is, the person's experience of mental pain, and the self-harming behavior following mental pain. As complete novices in brain imaging, it was a somewhat crazy idea to plan such a study. Fortunately, we got support from Ruth Lanius in Canada, the author of one of the

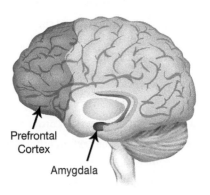

FIGURE 3.2. The amygdala, seat of the emotional alarm center, and the prefrontal cortex, seat of the rational brain.

Source: https://www.abqjournal.com/808912/retrain-your-brain-for-better-financial-decisions.html.

fMRI studies mentioned earlier, as well as from the Department of Neuroradiology in Bern. We decided to measure electrodermal activity while the patients' narratives were in the brain scanner, but we first had to get the necessary special equipment: the EDA recorder, headphones, etc., that would be compatible with the MRI machinery, which includes a brutally strong magnet that can immediately put out of order any normal EDA recorder. We ran many trials before things were functional, with the two of us acting as the guinea pigs in the MRI machinery. The MRI scanner was only available on Sundays, so we spent countless Sundays at the Department of Neuroradiology, together with our amazing patients, who had readily agreed come to the clinic on a Sunday and spend nearly an hour in the MRI scanner. Patients told us they were motivated to participate because they wanted to contribute to a better understanding of suicide, with the hope that this could eventually lead to effective prevention.

First, we fully transcribed each interview we had done previously and identified the sequences where patients expressed strong emotions related to the suicide attempt. We then selected sequences in which patients talked about the suicide action, such as taking pills, cutting, etc. We also recorded a neutral episode, where patients talked about getting up in the morning, preparing breakfast, and going to work. The fMRI technique allows scripts of only thirty seconds between the noisy scans. We therefore had to condense the scripts, and, in order to make sure that they were accurate, we revised the text together with our patients. We recorded each script on an audio recorder, but thirty seconds is extremely short—it was a real challenge to get it done in twenty-eight or twenty-nine seconds and not thirty-one.

Here is an example of a mental pain script: "It's his birthday, I call him but he says that he doesn't have time and that he will call back. I feel that he has another woman. I get more and more worked up, I start getting palpitations, get agitated, and in a panic, I am irritable, yelling at my children."

Example of a suicide action script: "It is Sunday afternoon and I am alone. I fetch the drugs I kept in a bag from my car, it is Nefidipine and Temesta, I write short letters to my children and my parents. I put them on the table, then I start pressing the tablets from the blister into my hand."

Example of a neutral script: "I am at work and it is time for the coffee break. I get up, leave my desk and go to the coffee machine. I make my choice, and when it is ready, I take the cup. I drink my coffee while I chat with my colleagues."

The whole procedure was quite a strain on patients—and on us, the investigators. Ten patients agreed to participate, but we had to exclude two of them because they developed panic attacks in the MRI scanner. With another patient I stayed in the MRI room, holding her hand during the scanning procedure.

What did we find after the mass of data was analyzed by a biostatistics specialist? The results showed that when patients talked about mental pain and emotional stress, there was reduced activation in a specific region of the PFC, the dorsolateral prefrontal cortex (DLPFC, Brodmann Area 46), and the medial prefrontal cortex (figures 3.3a and b). These regions of the PFC are related to working memory, self-reflection, appraisal of internally generated emotions, past and future events, and possible rewards—experiences we relate to our sense of agency.

Our interpretation of the results was that with the script-driven recall we had switched on the suicidal mode, which for our patients meant the deactivation of a PFC brain region associated with the ability to decide rationally how to deal with

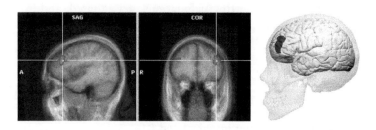

FIGURE 3.3A AND B. Brodmann area 46 of the PFC with areas of reduced neural activation.

Source: 3a: author; 3b: https://www.wikiwand.com/en/Brodmann_area_46.

emotional stress and mental pain.[5] We then looked at the brain activation of the suicide action scripts (buying drugs at the pharmacy, getting the pills out of the blisters, etc.), and we found a relative *increase* of neural activation in an area of the PFC (Brodmann Area 6), the so-called premotor cortex, which is associated with initiating movements and steering behavior. So, and I am aware that this is a very bold interpretation, the findings can be seen as supporting the theory that starting the suicide action may be a proactive way of escaping from the helplessness and from the experience of being trapped with unbearable pain. If we take this idea further, we could assume that a suicidal action, such as overdosing, could be recorded in the memory as "useful" behavior because it brings immediate relief and terminates the unbearable state of mental pain. This could be a plausible explanation for why after a suicide attempt the risk of future attempts (and of suicide) is massively increased and remains so over decades: The engram (the memory trace) of the suicidal action may be saved in the brain as a positive rather than negative experience, as one might expect.

  I am fully aware that there is a danger of overinterpreting the results of this small study in order to make it fit our concept of the suicidal mode. Studies measuring functional neural activation under certain conditions provide a confusing number of brain areas that light up with these nicely colored pictures of the brain. Statistical analyses are complex and leave a lot of room for interpretation.

In the stories of their suicide attempts some patients said: "This was not me acting like this, I was not myself, I felt as if I was controlled by an autopilot program." Surprisingly, this aspect of suicidal behavior is rarely mentioned in the professional literature. It is well known that high levels of emotional stress can lead to emotional numbing, to feeling detached from the body, and even to indifference to physical pain. We therefore asked our fMRI study participants to fill in a questionnaire, the Peritraumatic Dissociation Index (PDI), immediately after the initial narrative interview. They were asked to recall the acute suicidal crisis and rate the eight items on the PDI. Questions include: "Did you feel confused or disoriented? Did you feel numb? Did you have moments of losing track of what was going on—that is, did you 'blank out' or in some other way not feel you were part of the experience? Did you find yourself going on 'automatic pilot'? Did your sense of time change during the event—that is, did things seem unusually sped up or slowed down?"[6] We did expect scores that would indicate some dissociative symptoms, but we had not expected that the scores would be in the range or even higher than those found in studies with Vietnam veterans.[7] We concluded that the suicidal mode is frequently related to acute dissociative states. Dissociation may be another reason why a suicide action is usually not recorded in the brain as a negative experience. At that moment, people are often acting like robots, with numbed emotions. A dissociative autopilot mode as such is not a frightening experience. Analgesia (insensibility to pain) is another typical symptom of dissociation.

Here is what a thirty-year-old man told me: "I got into a kind of trance. I don't know how else one could call this state. Suddenly, one does not perceive anything what happens around"; "when I cut myself it did not hurt"; "I swallowed the pills without thinking much. I emptied the bottles and stuffed the pills in me"; "I did not realize it. I knew what I was doing. But I did not feel it. It did not hurt."

A twenty-year old female student said: "I was in my room and was cutting my arm deeper and deeper. I did not hear anything during the whole time. If somebody had come I would have cut further in my arm. I did not realize it. I just did not feel anything. The pain of the soul was the most dominant part. And I think that one can do this with a great soul pain only. The inner equilibrium was not there anymore."

Some suicidal dissociative states can last hours or even days. The concept of dissociation includes the notion that the self is made up of subsystems that can become disconnected.[8] The story of Mr. K. highlights this phenomenon.

Mr. K. age fifty-six, self-employed, had a history of unclear physical problems, which seriously impaired his ability to work. He was a dedicated skilled worker and was proud of what he had achieved. After two years of unsuccessful treatment with several medical specialists, he started to weigh the pros and cons of suicide. He made an appointment with the psychiatric outpatient clinic and was seen by the duty psychiatrist, who admitted him to the inpatient unit. He was discharged after six weeks of inpatient care. His wife suggested he give up his business, which she saw as the cause of his physical problems. During an argument with his wife, their twenty-eight-year-old daughter joined in to criticize him for having failed as a father because his business had always been his first priority. Being heavily criticized by his beloved daughter was "like a stab into my heart." He went out into the garden and lit a cigarette. He says that "it made a click" in his head, and he knew that the next day he would kill himself. He had dinner with his wife, watched TV, slept all night, had breakfast with his wife, then left, claiming he had some errands to run. He was totally calm. It was a cold winter, with some minus 10 degrees Celsius. He drove up to the mountains, parked his car in a forest, and took forty-two sleeping pills, together with a bottle of liquor. If that wasn't enough, he reasoned, the cold temperature would do the rest. He was found the next day, lying in the snow by somebody who had seen the empty car. He regained consciousness after two days in the hospital.

When later watching the video-recorded interview, he said: "I am shocked by how cold-blooded and indifferent I was. The moment I had made the decision I was like in a tunnel, no thoughts to the right, no thoughts to the left, it was like a PC where you press 'Enter' and then the program runs by itself."

Box 3.1: The suicidal mode and dissociation

The concept of the suicidal mode was introduced in cognitive-behavioral therapy by Aaron Beck and David Rudd.[9] A mode is defined as a response pattern to exceptional demands, incorporating cognitive, emotional, physiological, and behavioral systems of personality organization. Once established, a mode will be switched on by a typical triggering event. In suicide research, the term *suicidal mode* has proven particularly useful in understanding the sudden onset of suicidal mental states. For individuals with a history of attempted suicide, the threshold for activation of the suicidal mode will be lowered. Triggers can be either internal (e.g., by a thought, feeling, or image) or external (situations, places, people). *Cognition* in the suicidal mode is dominated by pervasive hopelessness, with the individual seeing no other solution than to put an end to an unbearable mental state.[10] The focus on long-term life goals is lost, as if the person was struck by a sudden mental myopia. *Emotionally*, the acute state is usually experienced as mental pain, often accompanied by anxiety, panic, embarrassment, humiliation, and self-hate. *Bodily functions* reflect an acute stress reaction with high inner arousal. The *behavioral system* is activated, experienced as an impulse for flight.

A suicidal mode typically has an on/off mechanism and can occur suddenly—for some patients seemingly out of the blue. The suicidal mode is time limited. One of the aims of therapy is to educate suicidal individuals so that they learn to recognize early warning signs and use specific skills to cope with a mental pain experience by engaging safety strategies.

Dissociation

In the diagnostic manual of the American Psychiatric Association (*DSM-5*), dissociative disorders are defined as a "disruption in the usually integrated functions of consciousness, memory, identity, or perception of the environment."[11] The disturbance may be sudden and the disruption partial. Typical symptoms of dissociative disorder include changes in the perception of one's self, one's senses, the environment, and time. These symptoms are transient and may vary greatly in severity. Dissociative symptoms occur in connection with acute trauma reaction, post-traumatic stress disorder, acute stress reactions, and, sometimes, panic attacks. Typical descriptions of dissociative symptoms are feeling confused or disoriented, feeling numb, acting in a trance or as if in a dream, functioning in the "autopilot mode," and not feeling connected to one's own body, with no sense of somatic pain (analgesia) and with an altered sense of time.

In the acute suicidal state of mind, dissociative symptoms such as emotional numbing, detachment from the body, and indifference to physical pain are frequent. Israel Orbach argued that mental pain and dissociation belong together: "If the natural protective shield against harm to the body gives way in the face of mental suffering, then the choice of self-destruction becomes very plausible. If one does not feel physical pain and is indifferent to the body, then it becomes easy to turn aggressively against one's body and one's physical existence."[12] In a dissociative state the suicidal person cannot reinterpret the past or foresee the future; they are "trapped in a seemingly unchangeable anguished present."[13]

I remember a workshop where the speaker presented a risk-factor model of suicide. Someone in the audience got up and said: "I am a surgeon; psychology is not really my thing. But can anyone explain to me how people with thoughts of suicide they may have had for weeks and months on the spur of a moment decide to go ahead and kill themselves, by shooting, jumping, or whatever?" I replied: "The answer is that usually some trigger—which may be seemingly insignificant event, like a text message—may be the last straw, which sets off an emotional response of pain, hurt, and stress that has a traumatic dimension. This mental state is experienced as unbearable, and the brain function switches to the alarm mode, in which problem solving is reduced to only one goal, to end—or escape from—this unbearable condition. Most people say that they were not their usual selves in this moment. It's called dissociation." The surgeon answered, "OK, thanks, I got it. This is helpful." The concept of altered brain function under extreme emotional stress makes sense not only to surgeons but to the wider public, and particularly to people who have attempted suicide.

Later, we started to hand out an educational text to patients, which described the concepts of mental pain, suicide triggers, the suicidal mode, and dissociation. When people came to the next therapy session, they often told us that it was an eye-opener for them, that for the first time they understood what had happened to them in the moment they harmed themselves. More than once a patient told me that the insight was so important to them that they showed the handout to their parents and grandparents.

Insight gained: The concept of the suicidal mode and the related changes in brain function are central elements in understanding suicidal behaviors, and they make sense to patients because they reflect their own experiences.

In the patients' stories we often heard that they experienced a sudden switch back from the trance-like, autopilot mode to normal thinking. For instance, the twenty-year-old student, when describing how she had cut herself in the forearm without experiencing pain, watching the blood make rings in the water in the sink, said: "And then, all of a sudden, the fear caught me, and I was not outside of myself anymore. I took a cloth and wrapped it around my bleeding arm. Then I took the phone and called my mother. I told her what I had done."

There are more drastic examples. In my home town, we have at least four bridges, each of them eighty meters or more over the river Aare. Some suicidal individuals survive the fall into the water. They usually tell us that the very moment they jumped they realized they had they had acted on the wrong decision. This is the typical description of the switching of the suicidal mode back to the "normal mode"—rather late, though. A prominent example is Kevin Hines, who in 2000 survived a suicide attempt by jumping from the Golden Gate Bridge:

> "I hit free fall, and that second, I knew what I did was a terrible mistake."[14]

What could this switch back to the normal mode mean in clinical practice? Let me give an example. A patient calls me in the middle of the night and tells me she is sitting at the kitchen table, with a large package of painkillers she is pushing out of the blisters. She has made up her mind—after yet another painful separation—to end her life. What do I do? Call the police or an ambulance, then inform the psychiatric unit that I am admitting an acutely suicidal patient? No. First of all, it is a good thing that in her crisis this lady has decided to call me (never mind the time of the night). Second, I ask her to tell me what happened. She says that she had a verbal fight with her boyfriend, that he left and that she is now desperate and convinced that this is the end of their relationship, and that this is the third time this happened to her and it is all her fault. She is such an awful person, unlovable, and everybody would be better off without her. She is in a state of arousal. After a few minutes, I tell her to get a glass of water and a tranquilizer I had prescribed to her for such moments, and I ask her to swallow the pill while I am with her on the phone. We then talk together for another five to ten minutes, and I give her an appointment for the next day. When I ask how she feels right now, she says, "I feel tired, and I think I'll go to bed now. I'll come to see you tomorrow at four." The level of stress has reduced. The suicidal crisis is over. The suicidal mode is switched off. Talking to me has reduced the level of stress into which this woman had spiraled. Taking a benzo may be an additional help, which allows her (and me!) to go back to bed sooner. Before she called me, she had experienced the physiological symptoms of a stress reaction, just as if she had been threatened by a wild animal. We'll look into other ways to switch off the suicidal mode in some of the following chapters.

Insights from research only make sense when they can be translated into therapy concepts. Here is the big question: How can we teach people that suicide is not a rational decision and that we all have the capacity to correct our plans and actions—even at the last minute? Research has shown that when developing an action plan, the brain usually generates competing plans in parallel. We can learn to monitor the action development and abandon one action plan in favor of another.

SUMMARY

In this chapter we have seen that the autonomous nervous system reacts dramatically to experiences of mental pain. We have also seen that with functional brain imaging it is possible to measure changes in brain function when people are asked to recall the suicidal crisis. Our study supports the idea that the suicidal mode is saved in the brain's circuitry as a response pattern that can be activated by experiences triggering an existential crisis. In view of therapy for people who attempt suicide, the question will be if and how we can get people to learn to respond differently to a threatening inner experience, *before* the suicidal mode is switched on.

But, before we continue with how we can influence suicidal behavior, we need to look at the seat—the hardware—of our inner world, that is, the brain.

4

THE BRAIN AND SUICIDE

The neuroscientist and Nobel Prize winner Eric R. Kandel caused an earthquake in psychiatry with an article published in 1998 in the *American Journal of Psychiatry*.[1] The title was "A New Intellectual Framework for Psychiatry." The key sentence: "All functions of mind reflect functions of the brain." In other words: Everything psychological is also (neuro)biological. At the Department of Psychiatry in Bern, we had heated discussions about this "new framework." How dare this Dr. Kandel reduce the mind to something so prosaic as neurons and synapses? After all, Kandel had developed his grand new theory from his studies on *Aplysia californica*, a marine snail, which has only some 20,000 nerve cells altogether—a far cry from the highly developed species of *Homo sapiens*, with some 86 billion (86,000,000,000) neurons. What makes *Aplysia californica* particularly interesting is that it has as few as twenty-four sensory and six motor neurons. *Aplysia* has a natural withdrawing response to sensory stimuli. Repeated electric shocks get the animal to *learn* to withdraw from even small stimuli such as a light touch; that is, the sensory nerve cells develop a long-term memory of response to this kind of stimuli.

Kandel and his team isolated single neurons (they are really big!) from *Aplysia*, put them into a culture dish, let them connect, then studied under the microscope how the cells developed a long-term memory. Today it is established that neural learning involves the synthesis of certain proteins. It also requires the growth of new synaptic connections between neurons, a process which takes some twelve to twenty-four hours.

Kandel, as a trained psychoanalyst, was in an ideal position to convince psychotherapists that emotions and cognitions are related to brain function—after all, Sigmund Freud, originally a neurologist, in his early works had speculated that there must be a neural basis for his psychological concepts ("Project for a Scientific Psychology"). Of course, at that time the neurosciences were far from able to shed light into the black box of the brain. In 2006, when I attended a small conference in Palo Alto, I found myself standing next to Kandel at a cafeteria counter. When I told him how much I had enjoyed his first article, "Psychotherapy and the Single Synapse" (what a provocative title!), published in 1979, he answered: "Well, of course, psychotherapy *is* a biological treatment, what else is it? How is it going to work if not through changes in the brain?"

> It is intriguing to suggest that insofar as psychotherapy is successful in bringing about substantive changes in behavior, it does so by producing alterations in gene expression that produce new structural changes in the brain. This obviously should also be true of psychopharmacological treatment.[2]

No question, Sigmund Freud would be excited.

All the learning experiences throughout our lives would not be possible without biological processes and the changes taking place in the wiring of the brain. Kandel: "We are who we are because of what we learn and what we remember."[3] In the last twenty years, brain research has exploded and has given us a wealth of fascinating insights into the brain. Today, children learn at school that the brain constantly changes and adapts throughout life, every day, probably every hour. This is called the brain's *plasticity*. We have learned that brain function involves the flow of neurotransmitters, the building of new synaptic connections (synaptogenesis), and the birth of new neurons (neurogenesis), which migrate to brain areas where they are needed. Every day, some seven hundred new brain cells are born in our brain. And all this is related to the amazing functioning of our genes (the "gene expression"). It is through the genes that the brain is able to respond and adapt to life and its challenges. This mechanism is called the *gene x environment* interaction. It is the basis of our being in the world, our biography, our personal needs and vulnerabilities—which, as we shall see later, may be related to suicide.

Box 4.1: About Neurons, Synapses, and Neurotransmitters

The cell bodies of neurons and the bush-like growth of dendrites were first described by the Spanish physician and anatomist Santiago Ramon y Cajal in 1888. Today, the number of neurons in the brain is estimated at over eighty billion, and each neuron has up to 200,000 connections, called synapses. This is an incredibly complex hardware, the scale of which is nearly impossible to imagine. Although we like to say that neurons are "wired together," this is not quite right. Messages are sent through the branches (dendrites) by electric impulses. These impulses are passed from one neuron to another via synapses, which are like tentacles meeting. The message crosses the cleft between the synapses with the help of neurotransmitters. These are released into the synaptic cleft from the first (presynaptic) neuron. The molecules of the neurotransmitters then dock on the specific receptors of the second (postsynaptic) neuron, setting off an electric impulse in this neuron, which, itself, may have thousands of other synaptic connections.

Me or the brain? In 1983, Benjamin Libet, an American physiologist, published another provocative study.[4] He wired study participants to an EEG, put them in front of a fast-moving clock, and instructed them to perform a simple motor task by flexing a finger whenever they wanted but to pay close attention to the time when they were first aware of the "urge to move." The time of the actual movement was recorded by measuring the electrical impulse in the muscles. Libet found that the time of decision to move was about 200 milliseconds *before* the finger movement. In the EEG, however, it was found that a "Readiness Potential" occurred much earlier, some 550 milliseconds before the movement. Libet's interpretation was that the brain made the decision some 350 milliseconds before the participants *consciously* decided to move the finger. If the interpretation was correct, it meant that the decision to act had been taken by the brain and that the idea of free will was an illusion. I won't go into the controversial discussions that followed these experiments. Interestingly, Libet tried to save the concept of free will with later experiments, arguing that "we" (i.e., free will) at least had the power to *stop* the impulse some 100 milliseconds before the action would take place. Altogether, the interpretation of these experiments is a thorny issue. What remains is that the experiment challenged our concept of free will.

Daniel M. Wegner, a professor of psychology at Harvard University, argued that the brain "plays a trick on us": "It usually seems that we consciously will our voluntary actions, but this is an illusion."[5] In Wegner's opinion, we are tricked into believing that "we" are the source of our actions and thoughts. The "we" is a virtual construct of our brain, and what we perceive as "us" is a virtual agent. If we continue along this line of thinking, the question pops up: "Who" is it who actually makes the decision when a person carries out a life-threatening, suicidal action?

Insight gained: Brain research questions the concept of free will.

We'll need to come back to this question later.

A main source for our sense of identity is autobiographical memory. We are who we are because of what we learn and what we remember. This becomes apparent when we tell stories about ourselves. Personal memories are stored in our synaptic connections, which involve practically all our brain areas. After all, *Homo sapiens* is quite a bit more complex than *Aplysia californica*! Our life histories and the related memory traces, conscious or unconscious, are crucial for the continuity of the sense of who we are. Memories and stories belong together: "It is through narrative that we create and recreate selfhood."[6] Narratives, suicidality, the sense of identity, and biographical vulnerabilities will become central issues later on in this book.

The first studies looking into the relationship between suicide and brain function came from the Karolinska Institute in Stockholm, Sweden. In the 1970s, Mary Åsberg and Lil Träskman measured the cerebrospinal fluid, which also surrounds the spinal cord, in patients who had attempted suicide. In a groundbreaking paper published in 1976, they reported that patients who had made violent and impulsive suicide attempts had lower concentrations of a molecule called 5-hydroxyindoleacetic acid (5-HIAA), a breakdown product of the neurotransmitter serotonin, than the other suicide attempters.[7] And a few years later, they reported that patients who had low 5-HIAA levels in the following years died more often by suicide. This was truly exciting. Obviously, serotonin function had something to do with suicide!

Serotonin is involved in the control of emotional impulses, particularly aggression. If serotonin concentration in our cerebrospinal fluid is low, we have difficulties regulating our emotional impulses—which appears to be related to an increased risk for dying by suicide. This was a fascinating insight. In the early 1990s, there was optimism that the new serotonin-specific antidepressants (Prozac being the first one), which increased the serotonin concentration in the frontal cortical regions of the brain, could effectively reduce suicide rates.

As we shall see later in this chapter, unfortunately, these promises have not been fulfilled. Serotonin has continued to be a main focus of biological suicide research. J. John Mann and his team at Columbia University in New York studied serotonin transmission in relation to suicidal behavior.[8] Serotonin is produced in a cluster of neurons in the brain stem called the raphe nuclei, and from there it is transported to various regions of the brain associated with stress response. In their postmortem studies of suicide victims, they found that there was a lack of serotonin in the prefrontal cortex (PFC), the brain region responsible for problem solving, finding alternative solutions, learning from past experiences, and integrating long-term perspectives. Interestingly, in the brains of many suicide victims, the researchers found an increase of serotonin receptors in the PFC, which has been interpreted as an attempt of the brain to compensate for the lack of serotonin arriving in this part of the cortex.

For the question what brain function has to do with suicide, we may conclude that one main factor is the reduced availability of serotonin in the PFC. Impaired PFC function is related to deficits in finding solutions to challenging life situations and to impaired regulation of emotional impulses. Both are highly relevant for suicidal behavior.

Other neurotransmitters relevant for suicidal behavior are involved in the regulation of the stress system. One of them is the noradrenergic system. Higher concentrations of norepinephrine (NE) in the prefrontal cortex have been associated with higher levels of aggression and with suicidal behavior.

Insight gained: Serotonin plays a role in emotion regulation and suicidal behavior.

There is no such thing as a suicide gene. Genes that are related to suicidal behavior are involved in a wide range of neural functions, such as learning, problem solving, decision making, emotion and stress regulation, etc. Serotonin transmission in the brain is related to *SERT*, the serotonin (5-HTT) transporter gene. The function of *SERT* itself is controlled by a so-called promotor gene (*5-HTTLPR*). There are two variants of the promotor gene in the population: a long and a short variant. As our genes come in pairs (alleles) from both parents, we either have a long-long, short-short, or short-long combination. In 2003, Caspi and colleagues from the Institute of Psychiatry in London published an article in *Science* in which they reported that people with the short-short combination were more prone to develop depression and suicidal thoughts and behaviors.[9] It meant that with this combination of the control gene, the neurotransmitter serotonin was less effective in maintaining emotional stability than the other combinations. However, the important point is that the more stressful life events people indicated, the more the three groups of gene combinations differed in relation to depression and suicidal behaviors. Although there are controversies in the scientific world about the interpretations of this and related studies, one thing we can clearly say is: Gene expression and thus brain function are shaped by life experiences.

We have earlier in this chapter come across the principle of *gene x environment* interaction. Here we need to introduce the concept of *epigenetics*, which describes a long-term programming of gene function as a result of what we have experienced in our lives, particularly during our developmental years. Early activation of the stress system through traumatic experiences may result in long-term changes of the function of genes ("gene expression") that are involved in stress regulation. Most prominent in research has been the glucocorticoid receptor (GR) gene in the hippocampus, which also plays a major role in the development of depressive disorders. Long-term impaired function of the GR leads to an overactive stress system (the hypothalamus-pituitary-adrenal [HPA] axis) and to more impulsive-aggressive traits. The main mechanism for the long-term programming of a gene is that early stress responses result in changes in the DNA. Unfortunately, this mechanism is difficult to undo. The same is true for SERT, the serotonin transporter gene: an epigenetic long-term change of this gene function caused by early stress experiences may lead to lifelong problems with emotion regulation.

It has been known for a long time that early traumatic childhood experiences have long-term effects on several aspects of brain function, including emotion regulation and the ability for introspection and problem solving. We also know that traumatic childhood experiences can increase suicide risk many times over. The concept of epigenetics has provided us with a fascinating biological correlate of the relationship between early life stress and its long-term effect on brain function. For me as a clinician trying to understand the suicidal patient in front of me, these genetic studies made a lot of sense given what I was hearing from patients in psychotherapy sessions. We'll see later that stressful and traumatic events play an important role in the patients' narratives of their suicidal crises and in the understanding of personal vulnerabilities, pain, and suicide triggers.

As a result of an evolutionary leap, the human cortex (the gray matter) has increased enormously in volume, and the most important part of the cortex is the prefrontal cortex (PFC). The PFC is involved in many everyday functions; it is the seat of our working memory, and it is where problem solving and action planning take place. We have seen that stress regulation and suicidal behavior are related to PFC function. People with a history of attempted suicide have reduced activity in the PFC and are more prone to impulsivity during decision making.[10]

Fabrice Jollant and his team assessed the decision-making abilities in participants who had a history of attempted suicide.[11] They found that these people had particular difficulties in learning from experience, quite similar to individuals with a lesion in a region of the PFC called the orbitofrontal cortex. Later, the same researchers added another piece of research. They showed participants in an MRI study a standardized set of facial expressions, ranging from happy faces to neutral to angry faces. Participants with past suicidal behavior responded strongly to angry faces, with a hyperactivity in several PFC areas, specifically in the so-called Brodmann Area 47, a brain area involved in the processing of anger. Psychotherapists have known for a long time that problems with self-esteem and hypersensitivity to disapproval are related to suicidal behavior. If we add the concept of epigenetics, we can see that these studies fit nicely into the emerging picture of the effect of early trauma, altered gene expression, impaired PFC function, impaired problem-solving capacities, impaired stress regulation, and suicide.

Stress regulation has been central to any species throughout evolution. The amygdala is the part of that brain that looks out for dangerous situations; it is our alarm center. If the amygdala perceives what could be a potential existential threat, it pulls the trigger of the so-called stress axis. Stress hormones are released. In humans, threats of an existential dimension can be physical (e.g., being taken hostage) or psychological (e.g., finding your partner with someone else in a compromising situation). Once the amygdala is triggered by external or internal threats, it can get adrenaline, our stress hormone for acute conditions, to shoot from zero to a hundred within seconds.

A patient told me the following story:

I was stopping at the traffic light close to a restaurant, when I suddenly saw my girlfriend with someone else sitting there, holding hands and looking deep into each others' eyes. I experienced a sudden pang of pain and an immediate stress response. The traffic light turned to green. But the brain had already changed into the stress mode. The rational brain was shut off. I couldn't think rationally. The head was spinning, the heart pounding. I was trembling. I had an immediate impulse to turn the wheel and crash head on with an oncoming truck. At the last moment, I was able to regain control. The heart was still pounding, but slowly, my brain started to work again.

With this man, the important point is that he had been betrayed by a former partner years ago. But the roots went back much further. His father had left the family for another woman when he was five. His mother became depressed and was unable to care for the children. What we can see with this story is that this man's amygdala, the alarm center, was conditioned for experiences of abandonment and deceit. In psychological terms, we call this *vulnerability*. It means that in the acute suicidal state, early and current pain experiences seem to fit together like a key in a lock. It is therefore important to see that the current experience is a *trigger*—or energizer—for a suicide action and not the *cause*.

Insight gained: Early painful experiences may lead to later vulnerability.

The next question is if and how we can influence or regulate acute stress. Brain research has given us some fascinating insights into the mechanisms of emotion regulation. The authors of a paper I have always admired got fifteen volunteers to participate in a study involving functional MRI brain scanning.[12] In the scanner, volunteers were shown two sets of color photos, one set of thirty-eight neutral images and the other with thirty-eight highly negative scenes. First, the participants were asked to view the photos and become aware of their related feelings; next, they were asked to reinterpret the second set of pictures so that it no longer elicited a negative emotional response, for instance by generating an interpretation of, or a story about, each photo that would explain negative events in a less negative way. The outcome was that the study participants were able to control their emotional reactions by using *cognitive reframing* of the emotional content. For instance, a picture of women crying outside a church could be seen as attending a wedding instead of a funeral. But the important finding was that the brain scans in the two conditions showed two distinctly different patterns of activation. The first ("attend") condition showed an increased activation of the amygdala (the alarm center) and the orbitofrontal cortex, a brain area associated with emotion processing. In the second ("reappraise") condition, however, more activation in the lateral part of the PFC (the area involved in cognitive functions like problem solving) and a reduced activation in the amygdala were found. This study nicely demonstrates that the way we interpret emotional content (a PFC function) influences the response of the emotional system (a function of the amygdala). The authors concluded: "We can change the way we feel by changing the way we think, thereby lessening the emotional consequences of an otherwise distressing experience."

Hamlet says: "There is nothing either good or bad, but thinking makes it so." Shakespeare, 1602.

Amazing.

Neurobiological research has not given us the cause(s) of suicide, and my guess is that it never will. Basically, what we are seeing are *correlates* of mental activities. Brain research cannot give us the explanation for why suicide happens. It cannot tell us what triggered the amygdala. But at least the brain is no more the "black box" it used to be half a century ago. The same is true for the diagnosis of depression. Clearly, the neurosciences have brought fascinating light into the biological mechanisms related to suicide, including depression.

Neurobiological studies of depression and suicide focus on—you will not be surprised—problems with regulation of the stress system, that is, problems in the regulation of the stress hormones adrenaline and cortisol (the glucocorticoids), which are released into the blood by the adrenal glands. The glucocorticoid receptor (GR) is responsible for the regulation of the stress hormones circulating in the body via a negative feedback loop, that is, for the downregulation of the hormone (the adrenocorticotropic hormone) that stimulates the adrenal gland to release adrenaline. In depression, the receptor does not do what it should do, namely, "downregulate" the stress hormones after a stress response. Impaired function of the GR leads to an overactive HPA axis (the hypothalamus-pituitary-adrenal axis) and thus to a long-term elevation of the stress hormones. High levels of stress hormones have all kind of negative effects in the body. They are responsible for hypertension, arteriosclerosis, heart attacks, etc. But they are also responsible for negative effects on brain function, resulting in a reduced growth of new synapses and lowered production of new neurons. Interestingly, there is evidence that antidepressants can counteract this effect; they can increase neurogenesis by activating the glucocorticoid receptor.

When we look at patterns of brain activation in depressed patients, we see a decreased activity in brain regions of the cortex, above all the PFC. Postmortem studies with individuals who had a long history of depression even found a loss of volume in the PFC. This volume reduction in the PFC correlates with the duration of untreated depression. And there is also a reduction of the amount of neural connection between the PFC and the emotional brain (the limbic system). Therefore, we can say that untreated depression can severely impair numerous brain functions, one of them being stress regulation—which, as we know, is related to suicide. Researchers followed over a three-year period forty-six adolescents who had been given a diagnosis of depression.[13] Seventeen adolescents attempted suicide during follow-up. The assessment had also found in these seventeen a reduced volume of PFC areas, which are known to be related to emotion regulation. The PFC continues developing up to the age of twenty-five. This means that an immature PFC is a contributing factor for the tendency to impulsive behavior—including suicide—in adolescence.

In clinical practice, the relationship between depression and suicide has been clear for a long time. Before the development of modern brain research, clinicians knew that most patients with severe depression had serious thoughts and plans of suicide. Consistent with this clinical experience, in the 1990s, "fight depression" campaigns mushroomed in Western countries. These campaigns aimed to increase the awareness for depression in the general population, increase help seeking, fostering treatment, and finally reduce suicidal behavior. And there were good reasons for this. For instance, the authors of a Finnish study collected all the available information about seventy-one individuals who had died by suicide and who had been diagnosed with a major depression.[14] Three-quarters had been in psychiatric treatment, but only 45 percent were in treatment at the time of death, most had not received any antidepressant treatment, and only 3 percent had received antidepressant medication in adequate doses. This, of course, implied that un- or undertreated depression was related to the death of these patients.

In the 1990s, Göran Isacsson, from the Karolinska Institute in Stockholm, assessed the blood concentrations of antidepressants in over five thousand suicides.[15] He found that most people who had died by suicide had not taken antidepressants before their death. Later, he concluded: "Suicide is a fatal outcome of psychiatric illness. It is a serious and demanding task for psychiatry to prevent it. Depression is the core of suicidality—its treatment is the cure."[16]

Isacsson compared international data on suicide and antidepressant prescriptions. He argued that a fivefold increase of prescriptions for antidepressants correlated with a 25 percent reduction of suicides. This looked very promising for suicide prevention. Drug companies started to invest in the development of new antidepressants. The expectation was that serotonin-specific drugs would be particularly effective in reducing suicide, also in view of the fact that the older antidepressants, the tricyclics, were often lethal when taken in an overdose. "A suicide-preventive programme based on adequate treatment of all depressed individuals in the population may require that at least 5–10% of the population take antidepressant medication (including long-term prophylaxis)."[17]

The pharmaceutical industry pushed the new antidepressants—the SSRIs, SNRIs, etc.—not only because they were clinically promising but because they were economically much more interesting than the old tricyclics. A huge new global market appeared to lie ahead, suggesting enormous economic growth and profit. I owned no shares in the pharmaceutical industry. But for my clinical work I wanted to be part of the new era of effective suicide prevention. I got involved in an international multicenter study to investigate the antisuicidal effect of paroxetine, one of the new SSRIs (selective serotonin reuptake inhibitor). The objective was to show that long-term treatment with a serotonin-specific drug would reduce suicidal behavior in patients treated after a suicidal crisis. I got the OK from the local ethical committee. We started to enroll patients who had attempted suicide and allocated them to a treatment group and a control group. Unfortunately, very soon some problems emerged. Typically, the patients' motivation to continue taking a drug over months and even years dropped once the suicidal crisis had been resolved. Another problem was that when patients were considered to be suicidal (that is, had suicidal thoughts), they had to be excluded from the study for ethical reasons, a necessary regulation in any drug trials with suicidal patients. The study was abandoned within one year.

Insight gained: Measuring the effect of medication on suicidal behavior is extremely difficult, considering the many other factors involved.

Over the years, suicide experts developed more and more doubts about the role of the new antidepressants in suicide prevention. Critical articles started to appear. Herman M. Van Praag, a professor of biological psychiatry in the Netherlands and the United States, a well-known proponent of antidepressant drug treatment, wrote:

> A major strategy in the treatment of depression, both major depression and dysthymia, is the prescription of antidepressant drugs. The use of those drugs has risen substantially over the years. In The Netherlands, for instance, the increase was 12% yearly over the past 4 years. Of course, antidepressants are also

used in other conditions, but depression still remains the main reason for their prescription. Rightly, then, one may expect suicide rates to have gone down, in proportion. This, however, did not happen.[18]

Later, Van Praag went even further with an article titled "A Stubborn Behaviour: The Failure of Antidepressants to Reduce Suicide Rates,"[19] and Diego De Leo, one of the authorities of suicide prevention and the longstanding editor of *Crisis*, wrote:

> Although antidepressants may be effective in the treatment of depressive symptoms, the current evidence does not suggest that they have an effect in reducing the risk of suicide attempts or completions. Antidepressants do not address the variety of psychosocial factors that are strongly related to suicide and depression. Improvement in psychosocial functioning is independent from and slower than improvement in depressive symptoms.[20]

We later ran a study where we compared patients who were on continuous antidepressant medication over at least one year after a suicide attempt with drug-free patients over the same period of time. Suicide ideation and reattempts decreased in both groups, but we could not find any effect of the medication.[21]

Several studies reported an *increase* of suicidal behavior in the first days after starting antidepressants. Some reports blamed the SSRIs, arguing that this group of antidepressants, at the beginning of treatment, often caused restlessness and agitation, which increased the likelihood of self-harming behavior (to be precise, they seemed to increase suicide ideas and attempts but not suicide deaths). These findings prompted the U.S. Food and Drug

Administration (FDA) in 2004 to order pharmaceutical companies to add a "black box warning" to the labeling of all antidepressants used for children, adolescents, and young adults. Later, in 2007, reports appeared that the decline in antidepressant prescriptions might be a reason for an increase in youth suicides.[22] Although there is no general agreement, many experts believe that the black box warning may have done more harm than good in drastically reducing the use of antidepressants in teenagers. The current practice now is that health professionals prescribing antidepressants to young patients must inform them and their families about the possible early increase of suicidality and for the first two to four weeks have to keep a close eye on these patients.

So, with today's knowledge, what can we realistically expect antidepressant treatment to do to reduce suicide?

Bertolote and colleagues from the Department of Mental Health and Substance Dependence of the WHO in Geneva took a worldwide view on the topic, assuming that antidepressants would be effective in treating depression in about 52 percent of cases (a figure taken from the *World Health Report 2001*) and that some 50 percent of people with depression were correctly diagnosed and treated (which I personally consider far too optimistic).[23] The impact on suicide rates would then be about 7.8 percent, which would reduce suicide rates from a world average of 15.1 per 100,000 to 13.9 per 100,000.

Quite sobering.

And, in addition, we need to consider that in clinical practice, treating depression is met by many obstacles. First, depression needs to come to medical attention before it can be treated. Some 50 percent of people who die by suicide do not seek professional help—predominantly men—or they only present with physical symptoms. Those who are in treatment have to be motivated to take a drug that takes two to four weeks to have an effect and that may have side effects. If the drug treatment helps patients recover from depression, they should continue for months or even years, because depressive episodes have a high risk of relapse. Unfortunately, many patients stop taking antidepressants once they feel better.

The fact that depression is a risk factor for suicide is generally undoubted, and so is the need for adequate treatment of affective disorders, which includes proper prescribing of antidepressant medication, following established guidelines (e.g., NHS Foundation Trust Guidelines).[24] In the media, antidepressants are often criticized for not being effective. The fact is, and this is rather surprising for clinicians who often see dramatic improvement in patients treated with antidepressants, that in clinical studies on depression it is not easy to demonstrate that antidepressants do better than placebo. Clinical trials differ from clinical practice, where drug effect and side effects are regularly reviewed and treatment adapted accordingly, which may include switching or combining medication. Another factor is that the placebo response is an inherent part of any drug treatment. In clinical trials where patients and doctors are very much aware that there is a 50 percent chance to get an inert substance, the placebo effect is literally wiped out.

Box 4.2: About the Placebo Effect in Treating Depression

Antidepressant prescribing (like any other drug treatment) is always accompanied by a strong placebo response. When we hear "placebo," we may think of dummy pills, where patients believe that they are getting an active drug. However, the placebo effect is an integral part of practically all kinds of treatment. Studies using brain imaging clearly show that the patient's expectation of getting relief with a treatment can trigger changes in neural activation. This effect, as you might expect, is related to the way the health professional interacts with the patient. The basic issue is the trust in the therapist and the quality of the therapeutic relationship. An empathic medical practitioner who takes time to understand the patient's needs, who can motivate the patient to start a drug treatment, but who also respects the patient's need for autonomy will have better results with any drug treatment. The patient's brain will be stimulated by the positive attitude toward the drug—a healing effect that by many professionals is considered to be larger than the effect of the chemical substance alone. Drug and placebo response come as a package. In a classical multicenter study in the United States, the authors compared drug treatment with psychotherapy for depression.[25] First, they found no significant difference in outcome between drugs and psychotherapy. However, when they studied the video-recorded sessions, using a rating scale to measure the quality of the therapeutic relationship (Vanderbilt Therapeutic Alliance Scale, or VTAS), they found that the outcome in the different treatment conditions was significantly related to the VTAS scores.[26] The authors concluded that for both treatment conditions "the therapeutic alliance is a common factor with significant influence on outcome."

A LOOK AT TWO INTERESTING COMPOUNDS

Before closing the overview of drugs that act on the brain, we need to look at two substances with remarkable evidence of the potential to effectively reduce suicide risk.

First of all, lithium. Lithium is an amazing compound. It is a chemical element (number 3 in the periodic table), a naturally occurring trace element that in low concentrations can be found in drinking water. The psychotropic effect of the salt lithium carbonate was discovered accidentally by John Cade, an Australian MD in 1948. Cade found that an injection of lithium had a calming effect on guinea pigs, whereupon he ingested lithium himself to ensure its safety in humans (!). It soon turned out that lithium was highly effective in reducing mania in manic depression (today, called bipolar disorder) and in reducing the mood swings of the disorder, which has a highly increased suicide risk. But what is most interesting in our context is that lithium has a strong antisuicidal effect—even in cases where the effect on the course of depression is poor. And what may be even more interesting is that there are studies from the United States, Japan, Taiwan, Austria, Lithuania, and other countries showing a relationship between natural lithium concentrations in the tap water and suicide rates.[27] Higher levels of lithium in drinking water are related to lower suicide rates. The lithium concentration in the tap water differs considerably even between districts of the same geographical areas. Generally, blood levels in individuals are one hundred times or more lower than those used in lithium treatment. There is only one possible conclusion: A simple salt like lithium has an effect on brain function—which underlines the fact that suicide is related to brain function.

There is a second interesting substance that can have a remarkable antisuicidal effect: ketamine. Ketamine is an anesthetic known for a long time. In the year 2000, the first reports appeared that ketamine had a rapid—that is, within hours of a single intravenous infusion—antidepressant effect on severe and treatment-resistant depression. More reports have since then appeared, showing that ketamine (or esketamine) is an extraordinary chemical substance.[28] It can dramatically reduce acute suicidality in patients with major depression in less than one hour. I don't think that to date anyone really knows the mechanism of this effect. Some studies found that the effect of ketamine could be attributed to its effect on the amygdala; others found a reduced functional connection between the limbic system and the prefrontal cortex. Ketamine requires a word of caution: It is not without side effects, some patients do not respond, and some experts have argued that the drug might create a "high" or euphoric state, which, if the drug was administered repeatedly, might result in a state of dependency.

SUMMARY

As Kandel wrote: "All functions of mind reflect functions of the brain." In this chapter I have taken you on a brief journey into brain research and how brain function is related to suicidal behavior. We have looked at the serotonin story and at genes, the software regulating brain function, and we have looked at the brain regions involved in problem solving and stress regulation. In essence, stress regulation is the common denominator of the many aspects of brain function related to suicide. We looked at depression as a disorder of brain function and at the ups and downs of antidepressant treatment in suicide prevention over the years. The conclusion is that the effect of drug treatment on suicide is limited. Still, the fact remains that mental disorders are risk factors for suicide, which must be recognized and treated adequately.

Neurobiological findings are interesting correlates of brain function and suicidal behavior. They may help us understand certain psychological mechanisms related to suicide. However, they do not explain the individual psychological experience of the suicidal person. The question is: How can we access the personal experience of the suicidal individual?

5

PROBLEMS OF COMMUNICATION IN MEDICAL CONSULTATION

In 2015, the *New York Times* ran a prominent article about Manny Bojorquez, twenty-seven, an Afghanistan veteran who, after deployment, had experienced eight suicides in his battalion and who himself became severely suicidal. Bojorquez was brave and decided to contact a hospital of the Veterans Administration. He had actually tried getting help at the VA once before, right after one of the funerals, but had walked out when he realized the counselor had not read his file. With each visit, it appeared to him that the professionals who are trained to make sense of what he was feeling understood it less than he did.

> Mr. Bojorquez tried the system one more time out of desperation. After the spate of suicides in 2014, he called and said he needed help. The V.A. had him see a psychologist and psychiatrist.
>
> He told them that he wanted therapy but no drugs. Too many friends had stories of bad reactions. One, Luis Rocha, had taken a photograph of all his pill bottles right before shooting himself.
>
> "We get it, no drugs," he recalled them saying. But on his way out, after scheduling a return appointment in two months, he was

handed a bag filled with bottles of pills. He calmly walked to his car, then screamed and pounded the steering wheel.

He wanted to get better, so he started taking the medications—an antidepressant, an anti-anxiety drug and a drug to help him sleep—but they made him feel worse, he said. His nightmares grew more vivid, his urge to kill himself more urgent.

After a few weeks, he flushed the pills down the toilet, determined to deal with his problems on his own.[1]

An obvious problem of communication between client and health professional. The usual medical and psychiatric approach to suicide is to see it as a consequence of psychiatric pathology, mostly affective disorders. In the medical world, psychiatrists are the experts in treating mental disorders. Suicide and attempted suicide as human phenomena do not fit into the traditional illness model.

Let me be clear: I am not questioning the relationship between psychiatric disorders as risk factors and suicide. As a clinician faced with a suicidal patient, it is important for me to do a proper psychiatric assessment and, above all, to recognize symptoms of depression or bipolar disorder and, if indicated, prescribe medication. Yet, surprisingly, there is only very limited evidence that the treatment of psychiatric disorders does indeed have an effect on suicide rates. We cannot escape the conclusion that the medical model has long ago reached its limits in reducing suicide. Campaigns to fight depression have been implemented in most Western countries, but the results have not been convincing. In relation to the high prevalence of depression (5 to 10 percent of the population over the course of their lives are affected by affective disorders),[2] the proportion of depressed patients dying by suicide is in fact very small; what's more, the effect of antidepressants is unclear.

Most people who develop suicidal thoughts and plans do not think that they need medical treatment—this is above all true for men. "Real Men, Real Depression" was a campaign launched by the National Institute of Mental Health in 2003. "Real men" have difficulties accepting a psychiatric diagnosis. "Real men" do not seek help. They keep suicidal thoughts to themselves. "Real men" do not show weakness, nor do they need help. In medical consultations, men are often not recognized as being depressed because they avoid admitting mental problems but instead present with physical problems or turn to alcohol and aggressive behavior. Wolfgang Rutz, the author of the Gotland study (chapter 2), called it the "male depressive syndrome."[3] Depression in men is often masked by atypical symptoms like irritability, anger attacks, hostile/aggressive/abusive behavior, and alcohol and/or drug abuse or equivalents such as workaholism, excessive exercising, etc. What this means is that there is an important gender factor in the treatment of depression. Professor Jules Angst, a psychiatrist and epidemiologist from Zürich, put it bluntly: "Women seek treatment, men drink and die by suicide." When I met Wolfgang Rutz at a conference, he said to me: "If you find out how we can get men to seek help you'll get the Nobel Prize!" Unfortunately, men don't seem to change, and my Nobel Prize today remains as far away as ever.

But, of course, we should not only speak of the male population. In our early study, where we asked female and male patients one year after a suicide attempt who could have helped them to change their mind before they harmed themselves, only 10 percent mentioned a medical practitioner. In a more recent study, Kuramoto-Crawford and others asked 8,400 individuals who in the past year had indicated that they had developed thoughts of suicide why they had not received any help.[4] Three-fourths did not feel they needed mental health treatment. Other reasons for not seeking help were the costs, not knowing where to turn to, and negative attitudes toward mental health services. Even in the days and hours before harming themselves, many patients consider suicide plans as something "private," as something they do not want or dare to share with others. They feel that the busy medical professional will not understand their inner struggles, their loss of self-respect and self-hate. The example in chapter 2 of the patient visiting the primary care physician, presenting with a banal physical problem, then shooting himself after having left the surgery with an ointment and an elastic bandage is typical. This man had made the decision before the medical consultation and had obviously brought a gun with him—but still, he went to the busy physician's surgery on a Saturday morning. The only interpretation is that he left a small opening in which to change his decision at the very last moment. But the life-saving meeting of two minds—the professional helper and the desperate person—did not happen.

I have heard other stories like this. From my early work with general practitioners I had found that, generally, patients don't talk about their suicidal thoughts, and doctors don't ask. The medical model does not provide a platform for patients and physicians to talk about suicidal thoughts and plans. Medical models of suicide, by their nature, tend to ignore an individual's very personal experience. Suicidal people are seen as patients with a disorder, for which the medical professional is the expert. Treatment modalities are decided according to the clinician's theoretical model, which can be found in the textbooks and in the scripts used for medical certification. Thus, the health professional's understanding of suicide does not meet the needs of the suicidal patient as a human individual with a personal, complex, and rich inner world. Undeniably, there is a serious communication problem between suicidal patients and health professionals. Not surprisingly, in consultation, suicidal patients rarely talk about their intentions, nor do they spontaneously mention past suicide attempts.

A major problem with the medical model is that it assumes that suicide is the endpoint of a pathological development, like in a somatic disorder. For example, diabetes has genetic and environmental causes; the pathophysiology (the development of pathology) of diabetes involves the malfunction of several hormones (i.e., insulin, glucagons, and growth hormone) and their effect on several organs, such as the eyes, the kidneys, the heart, the nervous system, etc. The illness's development can be monitored, and indicated treatments (diet, medication, insuline, etc.) can be implemented, depending on the progression of the illness. Suicide and suicidal behavior do not fit into this linear medical model. Suicide is *not* the result of a long-term linear development eventually culminating in suicidal behaviour. We have quite a good knowledge of the genetic and biological factors involved—risk factors for suicide—but, again, they do not explain the individual suicide. Neither does a diagnosis of depression.

Suicidal people and health professionals do not speak the same language. I call it a Tower of Babel syndrome ("They may not understand one another's speech" [Genesis 11:1–9]). Table 5.1 shows the problems of communication between the suicidal individual and the average health professional.

Adrian Treloar and Terence Pinfold took an early interest in the attitudes of suicidal patients toward the care they had received.[5] They gave questionnaires to patients who had been

Table 5.1. Communication problems

Suicidal Person	Health Professional
Experiences suicidal thoughts as "normal" and logical considering the actual inner personal experience (sense of "truth")	Understands suicidal thoughts as a symptom of a psychiatric disorder—and as pathology—requiring medical treatment
Does not feel that the condition needs medical treatment	Explores for symptoms of psychiatric disorder and works out a plan to treat the disorder (usually involving medication)
Fears the reaction of others—and treats suicidal thoughts as something "private" and secret	Asks about suicidal thoughts as part of the psychiatric assessment
May put suicidal plans to the test with hidden messages to significant others	Is satisfied with the patient's—cryptic or straightforward—denial
Is afraid that the health professional will admit to inpatient care	Immediately admits to psychiatric unit because of fears of litigation in case of a patient suicide

admitted to an acute medical unit following an act of self-harm. Patients were generally rather positive in their evaluation of hospital staff. However, some results are particularly interesting. Nurses were found to be most helpful, followed by social workers. Medical doctors and psychiatrists were both rated significantly lower on a five-point scale. Social workers and nurses had better listening skills than doctors. Asked what kind of follow-up they would choose, 40 percent mentioned a social worker, 38 percent a nurse, and 16 percent a psychiatrist. Not very flattering for us psychiatrists.

FIGURE 5.1. Medical professionals and suicidal patients speak different languages.

Dr. House will forgive me for his appearance as a typical doctor in a graphic model I use for teaching (figure 5.1).

―⚬⚬⚬―

Many patients, men in particular, find it difficult to talk to their physician about emotional problems. In their own view they have failed, and they hate themselves for this; they are vulnerable and fragile, driven by shame, and devoid of any protective shield. They fear stigmatization. Consequently, they choose to stay on safe ground by presenting with physical problems only. As long as we equate suicide with mental illness, a pathological condition that needs medical treatment, many people with suicidal tendencies will not turn to health professionals to seek help. And if they do, very often—too often!—their encounter with the health professional will be a frustrating experience, as we have seen.

Medically trained professionals have learned that when faced with an acutely suicidal patient they have to do a quick risk assessment and then decide what to do for the patient's (and their own) safety. Even when trying to understand the patient's reasons for attempting suicide, it is difficult for the physician in the emergency department not to actively develop all kinds of scenarios in the back of their mind on how to get the patient to a closed ward in a psychiatric institution, how to call the ambulance and the police, etc.

Building an early "ad hoc therapeutic relationship" with the suicidal patient is undoubtedly the most important means to reduce acute suicide risk. In teaching, I like to use a key scene from *Titanic*. Maybe you remember Rose (Kate Winslet) and Jack (Leonardo DiCaprio) when they first meet at the stern of the *Titanic*? Rose has climbed over the railing and is about to jump. Jack starts talking to her. She says, "Don't come near. You are disturbing me" (of course he does . . .). "Go away." He does keep his distance and answers: "I can't. I'm involved now. You let go and I'll have to jump after you." He takes off his shoes. She: "Don't be absurd, you'll be killed." He takes off his jacket and starts talking about his father, who used to go ice fishing, and how cold the water is in winter. A wonderful scene. Of course, my message is not that therapists should be prepared to jump with their patients! The message is that connecting with a person who is acutely suicidal by, first, respecting their autonomy ("don't come near") and, second, starting any nonthreatening conversation will most likely get the person out of the suicidal mode (see chapter 3).

Gilburt and others interviewed patients after their inpatient stays in psychiatric hospitals in London.[6] When participants talked about their experiences of inpatient treatment, they did so largely within the context of the people that they had encountered during their admission:

> Listening was rated highly by service users. The ability to listen was described as a characteristic of being human and service users who had the experience of being listened to described feeling respected. Conversely, one service user rated a whole service negatively because he felt the staff did not listen to him. Listeners who were open, nonjudgemental and not patronising were valued.

The importance of the listening ability, of course, is not something new. The Hungarian psychoanalyst Michael Balint (1896–1970), in *The Doctor, His Patient, and the Illness*, told doctors to focus their attention on what it is the patient is trying to convey at that particular time.

> The doctor must give the patient the opportunity to communicate. If the offer is accepted and a flash occurs in the patient which is responded by the doctor, the doctor is in duty bound to discipline himself to continue to observe the patient's and his own contributions to the therapeutic process—but not to run after secrets, which blunt his ability to observe what is there before his eyes.[7]

In a professional meeting in Vail, Colorado, where I presented the model of relating to suicidal patients we developed in Bern, Alex Sabo, MD, a dedicated psychiatrist from Massachusetts, got up and said: "What it really means is that we must relate to the *person in the patient*." I was stunned by this wonderful short formula. Indeed, this is the message! Joining patients in a nonauthoritarian, nonthreatening way and showing genuine interest in them as a person, a unique individual, will create the ideal, nonthreatening setting to listen and share in.

Insight gained: Good clinical practice means listening and relating to the person in the patient.

In workshops for primary care physicians I tell the participants they should take suicidal patients as human beings who have their inner reasons to consider suicide—and not as mentally ill individuals with pathological and dysfunctional thinking. I tell them that they should try to get out of their usual role of the medical expert who asks about symptoms of a mental disorder in order to reach a diagnosis. Instead, they should sit down with the patient, show genuine interest in the person, and try to understand how they got to the point of considering suicide. This is what David Jobes so convincingly describes as the *collaborative approach*—in contrast to the medical approach (see chapter 9). In chapter 13, I will argue that this is the only way to jointly assess suicide risk and decide on what the person needs to be safe for the next few hours, days, and months.

There is one last point I need to make. Let's assume the suicidal person was successfully admitted to a crisis intervention unit. Jim Rogers and Karen Soyka argued that the crisis intervention model tends to adopt a superficial view of a patient's suicidality, missing out on the patient's suicide history and on developing the necessary strategies and long-term measures for dealing with future dangerous situations:

> Without the ability to explore these past events in the context of a fully present human encounter, suicidal individuals have little opportunity for creating new meaning and to grow from the experience.

It is this collusion to avoid discussing those events that supports the taboo nature of suicide, allows past suicidal behaviors to continue to exert control over clients, contributes to the self-referent stigma of suicide, and makes it difficult for clients to learn and grow from their experiences. Thus, while current approaches may provide the illusion that we are able to keep clients alive (illusory in that it is not supportable in the data), we believe that we are, in fact, colluding with clients to keep their suicidal pasts a secret and expect them to constantly live in the shadow of suicide.[8]

There will be more about risk assessment and treatment options for suicidal patients in chapter 13.

SUMMARY

In this chapter we have looked at the problems of communication between the professionals and suicidal patients—a Tower of Babel syndrome. However, this is much more than a communication problem. There is a huge gap between the two worlds. The inner world of the suicidal person, on the one hand, the "one-size-fits-all," illness-based, theoretical world of the medical person, on the other. In my view, this gap is a major reason why suicidal patients drop out of treatment or don't seek help at all. However, I have also tried to point out that health professionals often are in a truly difficult position vis-à-vis the acutely suicidal patient. I am fully aware of the difficulties we have in establishing even a minimal therapeutic relationship in an emergency room setting. Yet, even in an emergency situation, finding a way to relate to the human being in the patient is clearly the recommended clinical approach.

BOX: THEORIES TO EXPLAIN SUICIDE

A Choice of Psychological Models

Sigmund Freud, the inventor of psychoanalysis, saw suicide as the result of the individual's unconscious wish to kill another person, an impulse, however, turned against oneself or, more precisely, against the psychological introject of the other person. I have always felt close to analysts, and I admire them for their enormous amount of wisdom about the human psyche, but in my practical work with patients in crisis I needed a less sophisticated and more direct approach to suicidal patients. You can't work with the concept of the unconscious in the emergency department.

Edwin S. Shneidman, the eminent American clinical psychologist and gifted therapist, and Norman L. Farberow are known for their analysis of suicide notes, in which they discovered that people who had died by suicide often displayed an inability to consider a situation from a different perspective and often viewed their choices in a black-and-white (either all good or all bad) pattern.[1] Shneidman later formulated the "ten commonalities" of suicide. These included elements such as emotional pain, frustrated psychological needs, constriction of alternatives, hopelessness, and helplessness. In "Suicide as Psychache: A Clinical Approach to Self-Destructive Behavior" he concluded that however we may categorize people in terms of psychiatric illness or socioeconomic stress, the final pathway to suicide is always individual and is characterized by

> hurt, anguish, soreness, aching, psychological pain in the psyche, the mind. It is intrinsically psychological—the pain of excessively felt shame, or guilt, or humiliation, or loneliness, or fear, or angst, or dread of growing old, or of dying badly, or whatever. When it occurs, its reality is introspectively undeniable. Suicide occurs when the *psychache* is deemed unbearable by that person at that moment.[2]

Israel Orbach, from the Bar-Ilan University in Israel, a dedicated clinical and research psychologist, focused his studies on mental pain, with Shneidman as his mentor. He developed a mental pain questionnaire, the OMMP. He was one of the few researchers who recognized the importance of *dissociation*, which typically is present in the acute suicidal crisis and characterized by an insensitivity toward physical pain and an indifference to the body.[3] Orbach was convinced that mental pain and dissociation were necessary preconditions to self-harming behavior.

Roy F. Baumeister, an American psychologist, published a paper in 1990 with the title "Suicide as Escape from Self," an absolutely brilliant paper—although my guess is that Baumeister probably didn't see many suicidal patients. Baumeister provided a convincing concept of the dramatic changes that happen in the suicidal mind. He stringently described the individual's urge to escape from aversive emotional states, which are the result of severe negative experiences that are incompatible with the person's expectations and needs and for which the individual blames themself. What follows is a *"deconstructed" mental state*: "Deconstruction removes meanings from awareness and thereby reduces actions to mere movements; as a result the internal objections vanish. . . . Death may come to seem preferable in the short run to one's emotional suffering and the painful awareness of oneself as deficient, and the long-range implications of death have ceased to be considered because of the extreme short-term focus."[4]

Mark G. Williams, a clinical psychologist at the University of Oxford, saw suicide as a form of *"arrested flight."*[5] In animal behavior literature, arrested flight describes a situation where an animal is defeated but cannot escape. The key element is the suicidal individual's perception of being helplessly trapped as a consequence of humiliation or rejection, a situation from which there appears to be no escape and no rescue. Williams and Pollock suggest that suicidal behavior (whether the outcome is life or death) should be seen as a "cry of pain" rather than a "cry for help."[6]

Ronald W. Maris, a professor of psychiatry and sociology from South Carolina, in 1981 published *Pathways to Suicide*, a book I have always admired. From his painstaking analysis of the suicidal developments in the lives of suicidal people, he concluded that suicide is related to the individual's entire *life experience*, from which it evolves as a possible solution during times of repeated crisis because of unsolvable difficulties, failures, or conflicts. The suicidal individual has "lost faith in his or her ability (or willingness) to manage being human." Suicide is understood as a development in the context of many interrelated factors, which include psychiatric, biological, psychological, and social circumstances ("suicidal careers"). In life situations where nonsuicidal alternatives are blocked or used up, suicide can emerge as a possible resolution of an unbearable adverse human condition.[7]

Aaron T. Beck, from the department of psychiatry, University of Pennsylvania, is *the* father figure of cognitive behavioral therapy (CBT). A key element in Beck's theory is the concept of *modes*, described as cognitive, affective, motivational, and behavioral schemas. They are typically activated as responses to adverse stimuli. M. David Rudd described the negative core beliefs related to the suicidal mode, which are about the self, others, and the future, with hopelessness as a key element.[8] Typical negative beliefs are: *I'm worthless, nobody really cares about me, things will never change, everyone else would be better off if I was dead.*[9] Thoughts that a negative situation is irreversible and the state of mind unbearable are characteristic for the suicidal mode. Once a person has acted on the negative thoughts with self-harming behavior, the suicidal mode will be deactivated but saved as a response pattern, which may be switched on any time in the future by a specific triggering experience. Modes, by definition, are time limited, but this knowledge is usually not accessible to a person in the acute suicidal crisis.

Marsha M. Linehan, an American clinical psychologist from the University of Washington, is the undisputed expert of the problems with emotion regulation, typical for people with a diagnosis of a borderline personality disorder (BPD), a diagnosis known to be difficult to treat. People with a BPD often learn to use self-inflicted injury (typically cutting) as a means of emotion regulation. BPD is strongly related to suicide attempts and suicide. Linehan's cognitive behavioral model assumes that the underlying problem is a learned, dysfunctional coping mechanism for acute emotional pain, related to a *deficit in a person's problem-solving skills*. Suicidal behavior doesn't arise from a single cause; biologically based deficits in problem-solving capacities, greater sensitivity to emotional stress, black-and-white thinking, etc. are some of the factors involved. Dialectical behavior therapy (DBT) has become the standard treatment for borderline personality disorders.[10] The "dialectic" in DBT refers to the tension between accepting the patient exactly as he or she is in any moment and simultaneously pushing the client toward changing behavioral patterns.

Thomas Joiner, a research psychologist at Florida State University, developed the interpersonal theory of suicide.[11] In this theory, individuals engage in serious suicidal behavior as a result of their perception of interpersonal experiences, above all a perceived burdensomeness (being a burden to others) and a thwarted belongingness (feelings of alienation). These feelings lead to the belief that one's death is worthwhile to others and to the desire to die. Repeated experiences of pain, often by self-injury, are seen as a way of acquiring *the capability for suicide*.[12]

As a summary of this brief review of psychological models of suicide, we can say that an acute and unbearable experience of psychological pain as a consequence of being deeply hurt, the inability to find a solution, an altered mental state (suicidal mode, dissociation), and negative beliefs about oneself, rooted in one's biography, are the main concepts we can take with us as we proceed with our journey though the suicidal mind.

A View on the Medical Model

Moving from psychological models to medical models means entering a different world. The term "medical model" stems from the notion that psychiatry is a medical discipline, primarily focused on the biological aspects of mental illness.[13] To understand this frame of thinking in relation to suicide, we must go back to the early classical studies that investigated the *frequency of mental disorders in relation to suicide*.[14] The general conclusion was that more than 90 percent of individuals fulfill the criteria for the diagnosis of a psychiatric disorder, with affective disorders being the main diagnosis. Other relevant diagnoses are personality disorders, substance use, and schizophrenia. More recently, within the diagnostic category of affective disorders, in addition to unipolar depression, bipolar disorder has been identified as a major suicide risk factor.

A basic problem in determining the role of mental disorders is that the relevant information on people who have died by suicide can only be gathered indirectly. The early studies relied on hospital records. Later studies used what is called *psychological autopsy.* This is a combination of interviews with those closest to the deceased and the collection of corroborating evidence such as hospital and medical practitioner records, social work reports, etc. A review of 154 psychological autopsy studies found psychiatric disorders in 91 percent of suicides in Western countries;[15] in Asia and in developing countries, this percentage appears to be much lower.

For over more than half a century the illness model was used to conceptualize suicide as a consequence of mental illness. Still today, the medical model largely determines how health professionals approach suicidal patients: Recognize and treat depression (or other psychiatric disorders), and you treat suicide risk. Research within the medical model is basically risk factor oriented, as is the case in somatic medicine. There are probably thousands of studies that have compared suicidal with nonsuicidal populations. This has led to an overestimation of the practical importance of single risk factors. Researchers and practitioners tend to forget that risk factors are no more than statistical measures. Depression increases the risk of dying by suicide, yet the vast majority of depressed patients will not die by suicide, and many will not even think of suicide. We have looked into the story of depression as a disorder of brain function in chapter 3. Risk factors are pieces of the mosaic, but unlike a puzzle, where we strive toward the whole picture, this puzzle will never be complete. Risk factor models will not help clinicians understand the very personal meaning of suicide for the patient just admitted to the psychiatric ward. And what's more, risk factors alone will not help the health professional predict the individual patient's suicide.

J. John Mann and his lab at Columbia University is the hotspot for biological suicide research. Mann launched the "stress-diathesis model":[16] Diathesis stands for a constitutional vulnerability to develop a disorder (suicide), such as impulsive and aggressive traits. Stressors can be internal (e.g., depression) or external (e.g., problems with relationships or employment). We have seen in chapter 3 that neurobiological theories of suicide include factors such as abnormalities of neurotransmitters (e.g., low serotonin availability in the PFC), genetic dispositions (e.g., familial dispositions and impulsive aggression), epigenetic mechanisms (caused by stressful childhood experiences), an overactive stress system, and functional and structural changes in the brain (reduced PFC activation, reduced connectivity between cortex and limbic system, etc.). The stress-diathesis model is more sophisticated than a simple list of risk factors, as it tries to accommodate the interaction between biological and psychosocial risk factors.[17]

Attempts to formulate comprehensive models of suicide usually result in complex graphic figures with lots of arrows indicating possible connections and interactions that finally lead to suicide. As a health professional with two hats, being practitioner and researcher, I can understand the need to explain suicide with a grand theoretical model. However, I also see the huge gap between theory and practice. Years ago I attended a workshop by Thomas Insel, the former Director of the National Institute of Mental Health (NIMH), an institution with a budget of over $1.4 billion. The title was something like "Suicide, from Genes to Molecules to Treatment and Prevention." The presentation had a compelling internal logic. According to Insel, before we can develop effective treatments the goal in mental health research must be to fully understand the pathogenesis (the development of a disease). When writing this chapter, I found an email I had written to Thomas Insel following the workshop:

> I spoke to you briefly in one of the breaks, sharing your vision about a new psychiatry. Indeed, I was impressed by the wealth of knowledge presented during the symposium, and I look forward to a new era of clinical practice. In the symposium, however, it sometimes seemed difficult to recognize the translational part of it. . . . I was a little bit worried by the call for a "mechanistic" psychiatry. If mechanistic means that treatments are more based on a scientific understanding of brain function, that's fine with me. However, contrary to Bob Stein's views, I hope that apart from becoming more scientific, psychiatry will continue to be a highly developed art of therapeutic interaction with human individuals.

Later, I stumbled over a remarkable blog in *Psychology Today* (posted May 23, 2017), which quotes Thomas Insel:

> I spent 13 years at NIMH really pushing on the neuroscience and genetics of mental disorders, and when I look back on that I realize that while I think I succeeded at getting lots of really cool papers published by cool scientists at fairly large costs—I think $20 billion—I don't think we moved the needle in reducing

suicide, reducing hospitalizations, improving recovery for the tens of millions of people who have mental illness.[18]

Still, many of my colleagues hope that one day we shall have an all-encompassing model that we can not only use to predict suicide risk in the single patient but also to cure patients from suicidality. In my view, this is a futile goal. Suicide—and this is one thing we have learned for sure—is an enormously complex phenomenon. What is missing in these theoretical models is the individual suicidal person.

Researchers deal with statistics and theoretical concepts, clinicians deal with real individuals. We must abandon the idea that even the most sophisticated model of suicide will eventually provide us with the miracle cure for suicidality.

6

SUICIDE IS NOT AN ILLNESS

A remarkable study caught my interest in the early nineties. I referred to it in my introduction to a presentation at the fourth European Symposium on Suicidal Behavior, in Odense, Denmark, in 1994:

> Harding et al. (1986)[1] in their studies on social values in Western Europe reported that Danish people thought suicide to be a more acceptable behavior than married men or women having an affair. So, in this country and maybe in most other Western countries the man or the woman from the street most probably believe that suicidal behavior is not a pathological form of behavior and that the decision to live or die by suicide is not primarily the responsibility of health authorities.[2]

Switzerland was not included in this study, but here, too, the prevalent opinion in my country was—and still is—that suicide is basically the personal and autonomous decision of every individual. I concluded that attitudes toward suicide are shaped by societal and cultural factors.

In a study briefly mentioned in chapter 2, I asked primary care physicians about their role in preventing suicide.[3] Here are the percentages of doctors agreeing with some of the statements presented to them.

Each individual has the right to kill himself:	58%
It is questionable to interfere with the decision of a person to kill himself:	15%
Suicide is always the expression of a mental disorder:	58%
Doctors can often prevent suicide:	57%

Interestingly, the practitioners endorsed the first and the third question with exactly the same percentage. It doesn't seem logical: People have the right to kill themselves, but at the same time suicide is the expression of a mental disorder, which gives medical professionals the right—or the duty—to interfere with the decision. Answers to statements 2 and 4 appear consistent with this.

In a later study, we asked sixty-six patients one year after their suicide attempt: "What had made you harm yourself?" Here are the answers:

Relationship problems/conflicts, loss:	33%
"Emotional crisis" (feeling low, depressed, etc.):	21%
Psychiatric illness:	15%
Emptiness, no sense in life:	8%
Physical illness:	1%
Other:	12%
Missing:	6%

We then asked, "Who should have taken some action to help you?" An astonishingly low 10 percent said a "medical doctor," some 20 percent said "relatives and friends," but the most amazing finding was that 52 percent said "nobody." The next question was: "Could you have accepted help?" 23 percent said yes; 50 percent said no.

Why did half of the patients say they could not have accepted help? A large survey in the United States found that of 8,400 adult respondents who had serious thoughts of suicide but were not in treatment, three-fourths did not feel that they needed treatment.[4] Could it mean that for these people suicide was a personal, maybe even a "normal," thing, something that in their view does not require medical treatment? Another question, of course, is how frequent thoughts about suicide are in a population. For instance, in Zürich, 20 percent of men and 24 percent of women aged thirty reported having had "persistent suicidal ideas" in their lives,[5] and in American high schools, an early study found that over 60 percent of the students reported having thought about suicide or attempting suicide.[6]

Insight gained: Thoughts about suicide throughout life are quite common, and many of those people do not think they need professional help.

So, the situation looks as follows: Thoughts about suicide are not at all a rare phenomenon. Yet the medical profession usually asserts that suicidal behavior (and this usually includes suicidal thoughts) is pathological and requires psychiatric treatment—if necessary, against a person's will. Things haven't changed much over the past thirty years. Susan Stefan, in an impressive volume on *Rational Suicide, Irrational Laws*, writes:

> Suicidal people are routinely sent to mental health professionals, most of whom equate suicidality with serious mental illness. This leads to inevitable conflict, and not only between treater and patient. There are huge social and economic costs to conceptualizing suicide and suicide attempts as inevitably the result of untreated or inadequately treated mental illness. Research shows

that, in general, mental health treatment for suicidal people does not work as well as treatment focused specifically at their suicidality.[7]

Stefan quotes a patient:

> As soon as you say something that hints you might hurt yourself, you have no control over what happens to you, this completely keeps me from going to a mental health professional. We need an atmosphere in the mental health field, where the response could be not one of panic but of support and understanding that someone is really hurting, and trying to help in whatever way the individual seeking help is willing to do.

Let me return to my presentation at the 1994 European Symposium:

> In my view the problem we have with understanding attempted suicide may be that (a) it does not fit the traditional medical model, (b) it is often triggered by interpersonal conflicts, and, (3) most medical doctors are not trained to give adequate professional help to people in acute emotional crises. I believe that most suicide attempts and many completed suicides are not so much the expression of a psychiatric illness but rather an expression of what Caplan in 1961 called an emotional crisis." . . ."If we work according to one model only we are likely to be blind in one eye and miss crucial aspects of behavior. The health professional who does not recognize a state of depression in an adult person will make a severe mistake in ignoring the need of prescribing antidepressants. The physician who does not understand the emotional state of a young patient unable to cope with an acute interpersonal conflict will not be able to help this patient. I do not want to challenge the classical studies showing that over 90% of people

who kill themselves had a psychiatric diagnosis. However, one must doubt the practical usefulness for the treatment of suicidal patients when in a review article of a leading medical journal one finds a total of 40 clinical, social and biological risk factors.

In my search for useful working models, I came across Ronald Maris's book (see Box: Theories to Explain Suicide). The key conclusion was that suicide can never be completely explained by acute, situational factors; instead, suicidality develops over time and in the context of certain social, psychological, and genetic conditions. In Maris's model, suicide is seen as a problem-solving behavior in people with difficult life histories, leading to self-destructive behavior as a means of escape from a long accumulation of painful life experiences. "No one suicides in a biographical vacuum; life histories are always relevant to the final act of suicide."[8]

In my work as a psychotherapist, I had learned how important it *is to understand the patients' problems in a biographical context.* To illustrate this, let's look at two psychotherapy cases, which, for a change, are not about suicide.

CASE VIGNETTE 1: THE CULTIVATED ITALIAN

After eight months of inpatient treatment for depression, a middle-aged Italian man was referred to me for outpatient care. The referring colleague said to me on the phone: "Good luck! A hopeless case. Treatment-resistant depression. We have tried everything." In the first session I was pleasantly surprised by this man's soft and cultivated appearance. I continued the heavy combination of antidepressant drugs for some six months, without any signs of improvement. I then decided to let him tell me about his family background, his childhood in a village in Calabria, in southern Italy. What came out was that two years ago, after the death of his mother, the mayor of the village had somehow managed to seize the family house by "losing track of" the mother's will, which had been deposited with the mayor. The patient and his sister appealed to the local court but lost. In the following session, he hesitantly confessed to me that in fact he had made plans to drive to Calabria and to the village, and that he had told a friend of his to acquire a gun for him so that he could murder the mayor. Wife and sister eventually managed to get him to abandon his plan—which was when his depression began. His temper and murderous impulses stood in total contrast to the cultivated and soft manner he had adopted, and it was obvious to me that for him this was a highly shameful issue. After he had deposited his shameful confession that there was a hot-blooded southern Italian in him and that he could easily have ended up a murderer, his condition steadily improved. Six months later I started to slowly take him off the antidepressant medication, and he went back to work and resumed his normal life. I met him years later in the street, and he proudly told me that he was fine. He had restored his sense of identity as a cultivated and trustworthy person, the way his family and friends knew him. But we both knew that this was not the whole truth about him.

CASE VIGNETTE 2:
THE SWISS ACCOUNTANT

The fifty-eight-year-old accountant had been employed by a Swiss firm that got into legal trouble in the United States, being accused of tax fraud. Mr. H. came to me after several unsuccessful in-patient treatments, with the diagnosis of treatment-resistant OCD. He had developed a compulsion of repeating the content of TV news and newspaper articles for hours, because—as he said—he had lost confidence in his brain function, therefore he compulsively—and practically endlessly—had to test his memory. I was interested to hear when and how his troubles had started. It emerged that his mental problems began a few months after he had returned from New York, where he had to represent his company in court. He managed to get the company out of legal trouble—by lying to the court. In his superiors' view, he had fulfilled their expectations—mission accomplished. But Mr. H. became more and more depressed and withdrawn, started to ruminate, and was admitted to a psychiatric hospital, where he was "filled up with drugs." He eventually lost his job. I guess it would be more accurate to say he was dumped. Neither the CEO nor any other superiors had openly acknowledged what he had done for the firm or had contacted him when he was on sick leave. In the outpatient sessions he eventually opened up and told me about the high ethical standards he had learned from his father ("never lie, and always be reliable in fulfilling your duties"). Now we could relate his psychological breakdown and his problems with memory to his inner values, which got him into an unsolvable conflict. I introduced the "Catch-22" situation to him: A dilemma from which there is no escape because of mutually conflicting conditions. Never lie. Be reliable and fulfill your duties. After this therapeutic disentangling of the serious conflict of personal values that were such an important part of his identity, we reduced his medication, and with the help of behavioral strategies he eventually got rid of his compulsive ruminations. He kept in touch with me for years after termination of treatment, telling me how well he was, from time to time sending me treasures from his wine cellar.

In my work as psychotherapist I have learned that severe emotional crises are related to peoples' personal values and biographies and that effective therapeutic work requires that therapist and patient find a shared understanding of the background story.

Insight gained: Emotional crises have a biographical background. Let's see how we can relate this to suicide.

I shall never forget the rainy day at the clinic in Bern, in September 1995. During a coffee break, my colleague and friend Ladislav Valach made a most provocative remark: "You medical people have learned to watch out for signs and symptoms of pathology, to make diagnoses and treat disorders accordingly. But, suicide is not an illness—it is an action, and you medics have never learned to understand how people come to act in this or that way." Little did I know that this conversation would be the beginning of a journey into the suicidal mind and some twenty years later would result in one of the most effective treatments for people who have survived a suicidal crisis. Ladislav, a man of theories and research, gifted with a wonderfully creative and original mind, had a background in social and health psychology. His tiny office was crammed with printouts of—at least to me—incomprehensible data outputs, so that when you went to see him, you first had to locate him behind stacks of paper.

The psychologist telling me, a medical person, that I hadn't got it right with suicidal patients? A key competence of a psychiatrist. And, hadn't I conducted an in-depth study of one hundred cases, showing that depressive symptoms correlated with the severity of suicidal behavior? Ladislav was not a person to be put off easily. He started to tell me about something called action theory (which I had never heard of, nor had any other psychiatrist) and that all our actions are goal directed. But how could this make sense for suicide? I was convinced that suicide, above all, was a form of pathology, a breakdown of a person's rational thinking, similar to a PC crashing with a frozen screen. It just didn't make sense that suicide should be a form of goal-directed behavior. Death as a goal? How could one's own death be a reasonable goal? Still, I finally succumbed to the insistent and gently coercive charm of my colleague and agreed to write a case study of a thirty-six-year-old patient of mine who had come to see me after a suicide attempt but had dropped out of treatment after two sessions—and who ended his life by suicide a year later.[9]

George grew up in Zürich as the son of a professor of history whose life principle had been that with a strong will you can achieve everything. At school George could not fulfill the expectations of his father, who repeatedly told him that he was a failure and that he would never be good for anything in his life. At university he tried several subjects and finally chose music as his major. He married a woman who did not meet his parents' expectations. The couple had three children and developed a caring and warm family atmosphere—different from what George had experienced at home. He left his studies and started to build string instruments—violins. His violins were of outstanding quality, but he was not a salesperson. The business did not do well, and his wife had to earn the money to support the family.

Gradually, George lost self-confidence and became depressed and mentally paralyzed, unable to work. Still, he made the effort to get officially credentialed as a violin maker, which he thought would help him improve his business and support his family. He was turned down because he had not fulfilled the training requirements. Following this blow, he had extreme aggressive outbursts, repeatedly wrecking parts of the household. He made a serious suicide attempt. He was admitted to a psychiatric hospital, where he was discharged two weeks later. Two days later, in an aggressive outburst he demolished all the instruments in his workshop. He seized the car keys from his wife and left the house. He broke into the surgery of the local family doctor, a personal friend of his, got hold of two bottles of barbiturates, and drove into the forest. His body was found the next day.

One year before his suicide George had written a letter, which his wife found after his death:

> Nothing helps. I have tried everything in my life. Really, I am looking forward to make my exit. It is important to see that my life is a succession of failures in every respect. I messed up the relationships with most of my relatives, and I am in the train of doing the same thing with my wife. Everything wrong, always, always, always. My best wishes to everybody, including me. I do not know where the journey will go, but at least it does not stay in this vale of tears.

In the context of Ladislav's action theory, the case of the violin maker became an eye-opener for me. This suicide only made sense in the context of George's biography. It was obvious: His important life goals were related to the negative and painful experience of being rejected by his father. The need to be successful in his chosen way of life, to prove his father wrong, however, was coupled with a vulnerability to failure and rejection.

In Ladislav's words:

> The relevant life career aspects include the failure to achieve certain intended personal goals (work career) and family goals (relationship career, or "joint projects"). More specifically, these goals

were (a) to support his family through his work, and (b) to maintain a good marital relationship. In relation to these goals his self-evaluation was: "I am a failure (as his father had predicted), I am useless, and life is hopeless." Suicide emerged as an alternative goal to his original life-goals.

We concluded that when existential life goals are seriously threatened, suicide may become a goal. "To end a bad story," as Ladislav put it. Life goals or needs may be conscious or unconscious. We may strive to be successful, respected, admired. We may be searching for love. We may do everything to feel secure in a relationship. We may find fulfillment in caring for others. And so on.

Highly charged needs and personal vulnerabilities belong together like the two sides of the same coin. George's existential goal was to prove to his father, and to himself, that he was able to lead a successful life in his own right. Failure to achieve this goal was the related personal vulnerability, and it echoed exactly what he had heard from his father: "You will never be good at anything. As my son, you are a failure." George's life-oriented goals turned into death-oriented goals: "I do not know where the journey will go, but at least it does not stay in this vale of tears."

Insight gained: To understand the suicidal person, we need an understanding of the related biographical issues and important life-goals.

In the conclusions of our case study we stressed the importance of the biographical context. This matched well with what Maris had written: "Suicidal hopelessness is directly related to repeated depression, repeated life failure, and prolonged negative interaction and social isolation."[10] I met Ronald Maris at a congress and told him that our clinical studies supported his concept of pathways to suicide, or "suicide careers." He told me that, originally, he wanted to use the term "suicide careers" for the title of his book but then decided against it because it could be misunderstood, as readers wouldn't understand the meaning of "career" in this context.

In my medical training I had learned to take the lead in interviewing patients and to ask questions to find out what was wrong. From my early training in Oxford, mentioned earlier, I remember an important clinical experience. I had learned to use the Present State Examination, an extensive research questionnaire with 140 questions, developed to provide in-depth information on a patient's signs and symptoms. In the training group, one of my colleagues interviewed a seventy-one-year-old female patient. He had gone through all 140 questions with her, without finding much pathology. The last question was: "Has there been something else recently that we have missed?" The lady answered: "Yes. I think you should know that I am pregnant and that I shall give birth to Jesus Christ."

Insight gained: In a clinical interview, when you ask questions you only get answers. The interviewer will most likely miss crucial and personal issues.

Since then, the question of how to conduct good interviews with patients has accompanied me throughout my professional life. We'll see later why this is particularly important for interviewing suicidal patients. We explain our actions with stories. By explaining how we came to act in this or that way we give meaning to events so that the listener can follow the logic of our story. There is more to it: Storytelling helps us deal with difficulties we encounter in our lives. I have always loved the following sentence:

> When we are able to formulate the right story, and it is heard in the right way by the right listener, we are able to deal more effectively with the experience.[11]

A story told to an attentive listener is called a narrative. Narratives can be oral or written. Narratives—even if we only have an imagined listener—are crucial for the continuity of our sense of identity, creating an ongoing thread between past and present. But, and this is crucial, narratives create a bond between narrator and listener. The therapist becomes an ally of the patient, supporting the patient in the struggle against life's adversities. Patient and therapist enter a therapeutic relationship, that is, a working relationship between patient and a trusted professional. A therapeutic relationship built on trust, empathy, and understanding is a precondition for an effective therapy. A related conceptual framework is provided by John Bowlby's *attachment theory* and, in particular, by the concept of the *secure base*.[12] Bowlby's theory has always had a special meaning for me. In 1979, as a trainee in psychiatry, I attended a two-day seminar on attachment theory at Thorney Creek House, a Victorian manor house in Cambridge. The first morning, a pleasant elderly man with a cute little rucksack talked to me on the way to breakfast and asked me where I came from. Only when the seminar started did I realize that I had just met the great John Bowlby himself. Bowlby's attachment theory still today is *the* developmental theory, that is, the theory that explains the importance of attachment figures (parents) for the development of a child. The growing child needs parental figures that provide emotional security and responsiveness to the needs of the child. The experience of a secure base enables the child to develop trust and to explore the world. The concept offers itself to be transferred to the therapeutic relationship in general. A good therapist provides the patient with a secure base experience, which can be compared to the qualities of good mothering (or parenting), that is, to be "reliable, attentive, and sympathetically responsive, and to see and feel the world through the patient's eyes."[13] The therapist joins and encourages the patient in exploring "the various unhappy and painful aspects of his life, past and present, many of which he finds it difficult or perhaps impossible to think about and reconsider without a trusted companion to provide support, encouragement, sympathy, and, on occasion, guidance." The therapist accepts and respects the patient "as a fellow human being in trouble and that his over-riding concern is to promote the patient's welfare by all means at his disposal."

A man whose work I have always admired is Jeremy Holmes, a consultant psychiatrist from Devon, a lecturer at the University of Exeter, and a wonderful expert on attachment theory. His book *The Search for the Secure Base* should be a must on the reading list for every psychotherapist to be. The narrative approach and the secure base concept in psychotherapy are intrinsically connected. Narratives are the means for therapist and patient to find a joint understanding. The therapist's task is much more than mere listening. It includes empathic responsiveness and emotional proximity. With the support of the therapist, a story can be revised in the light of new experience, new memories, and new meanings. "Like an attuned parent, the effective therapist will intuitively sense when the patient needs stimulus and direction to keep the thread of narrative alive."[14] The good therapist will be highly sensitive to the patient's needs for autonomy and for guidance—like a good parent who provides the child with a secure base.

So, we can say that a good therapist provides to patients the conditions which allow them to change. This means that, basically, my task as a therapist is not to change the patient but to provide a secure base from which patients can find new insights, explore new possibilities, and mobilize the potential for personal growth. In my training years I had learned that "change is always self-change." Drawing from Bowlby's attachment theory, I have always had a strong belief and trust in my patient's own developmental capacities. The case descriptions of the cultivated Italian and the Swiss accountant earlier on in this chapter may serve as examples.

Insight gained: In a trustful therapeutic relationship, patients can mobilize their capacities to change.

The narrative approach was new to me and turned out to be a real challenge. I had to learn to relax, to stop myself from asking questions, and to trust the patients' capacity to give me a coherent story of how they came to the point of harming themselves. To my surprise, most patients spontaneously put their suicidal crisis into a biographical context. They were not merely talking about the actual emotional crisis but took me back in their biography and told me about their repeated struggles with adversities in their childhood, painful experiences, rejection, problems with self-esteem, self-blame, guilt, and many other unmet needs.

Remember Eric Kandel in chapter 3: "All functions of mind reflect functions of the brain." The act of storytelling has been studied with brain imaging techniques. You will not be surprised to hear that, among the many brain regions involved in storytelling, the frontal cortex (PFC) plays a particularly important role. Stories told to an attentive listener require story organization; that is, the storyteller needs to distinguish between story-significant and story-insignificant elements. In telling our story we need to access autobiographical memory and retrieve what makes sense in producing a consistent narrative—the meaning of which can then be grasped by the listener (therapist). Narratives are always reconstructions of the past; they are "true" for the teller, they are individual, and they give meaning to events to produce a coherent story. And they give meaning to our lives. In the wonderful *Handbook of Narrative and Psychotherapy*,[15] Jerome Bruner writes: "It is through narrative that we create selfhood (and) self is a product of our telling" and that "individuals who have lost the ability to construct narratives have lost their selves."[16]

The ability to tell a coherent story is called *narrative competence*. When I started with my narrative approach with patients, some patients began with the early years in their lives, and I got worried, thinking they would take the whole day to tell their story. But I was wrong. A story about suicide is not the same as a life story. There are innumerable stories we tell one another. There may be relationship stories, family stories, work stories, love stories, etc. In a story about a suicidal crisis—the suicide story—the narrator naturally concentrates on the elements related to the existential crisis. Many of these are rooted in early traumatic experiences (*trauma* comes from Greek and means "injury or wound"). As we saw in chapter 3, psychological injuries in childhood can have a long-term effect on the brain's response to emotional stress and may result in lifelong problems with emotion regulation and suicidal behavior.

Insight gained: The narrative interview with the suicidal patient is the royal road to a shared understanding.

Personal narratives full of hurt, pain, and shame require trust in the listener. Right from the beginning of using the narrative approach, I was genuinely touched by the trust people showed when telling me their stories. Still today, after hundreds of narratives I have heard, I feel extremely privileged that as a health professional I can take the time and support patients in telling their stories in a way they had never done before. After a TV interview, I received a ten-page letter from an eighty-four-year-old woman. She wrote: "I saw that you are a person who likes stories. Here is my story. You don't need to answer. It's just good to know that someone reads it."

To engage a suicidal person in therapy, we need an "active and purposeful collaboration between patient and therapist."[17] It is the foundation on which patient and therapist together can work toward a shared goal. Active treatment engagement refers to "being committed to the therapeutic process and being an active participant in a collaborative relationship with a therapist to work to improve one's condition."[18] Collaborative self-reflection is crucial in assisting the patient to reach his or her goals.[19] This is very different from the medical model, in which—like in somatic medicine—patients tend to be seen as passive recipients of a medical treatment (being told what is wrong with them and receiving a prescription for drugs), and doctors are the active, knowledgeable authorities.

Insight gained: In therapy we want patients who are active participants.

During a panel discussion I argued that most of the treatments for which there is scientific evidence that they can effectively reduce the risk of suicidal behavior have not been developed from theories explaining suicide but from general psychotherapy research and experience. They must be patient centered, collaborative, and foster the patient's engagement with the treatment. The main established therapy programs for suicidal patients are cognitive behavioral therapy (CBT), brief CBT (BCBT) for suicide risk, dialectical behavioral therapy (DBT), the collaborative assessment and management of suicidality (CAMS), and the Attempted Suicide Short Intervention Program (ASSIP). The latter will be introduced later in this book.

But what exactly is the therapy goal? To "cure" patients from suicidality? I don't think this is a realistic goal, particularly for people with a history of attempted suicide. Amazingly, even cognitive therapies are not very effective in reducing suicidal ideation, although they reduce suicidal behavior. A more realistic therapeutic goal is for patients to learn how and when to interfere with the suicidal development before the dangerous suicidal mode—the suicidal autopilot (see chapter 5) takes over.

Are psychological treatment models compatible with the medical model of suicide? Yes, they are. The suicidal patient needs a health professional who first of all is an attentive listener, helping the patient tell the story of what happened, to try to understand collaboratively with the patient the background and what triggered the suicidal crisis. But the suicidal patient also needs a clinician who makes a psychiatric diagnostic assessment. If there are symptoms of a psychiatric disorder, the clinician is responsible for adequate treatment. Severe depression is a risk factor for suicide and must be treated adequately. Narrative interviewing and the treatment of medical risk factors should go hand in hand.

As we have seen earlier, it is not the depression that kills but the person who acts on a suicidal impulse.

SUMMARY

In this chapter I have introduced a model of suicide that understands suicide as an action. Actions have a history, and patients have their own logical reasons for their actions. Most suicide-related stories include emotionally highly charged life goals. On the other side of the coin, however, are the related personal vulnerabilities. With the case study of the violin maker we have seen that the painful experience of being turned down as a certified violin maker jeopardized his existential life goal and triggered self-aggressive and suicidal impulses.

To access the suicidal person's inner world, health professionals need to be attentive listeners. Different from the usual medical professional's role, the attentive listener is in the *not-knowing position*. The task is to provide a secure base experience to the patient, that is, a therapeutic attitude characterized by a supportive and empathic attitude. Only then can a meaningful and shared understanding develop—an absolute prerequisite for a therapeutic and effective working relationship.

7

THE FRAGILE SENSE OF WHO WE ARE

Ernst Ludwig Kirchner, the famous German painter, spent his last twenty years in Davos, Switzerland, and shot himself on June 15, 1938. Eighty years later, I ran a workshop in Davos for primary care physicians. In the main street there is the wonderful Kirchner Gallery. Kirchner's expressionist paintings, with their strikingly and wonderfully strong colors, can be found in picture galleries throughout the world. Ernst Ludwig Kirchner was born in 1880. He studied architecture in Dresden but soon started to paint, moved to Berlin, and became a member of "die Brücke," a group of avant-gardist painters in Germany. He saw himself as "one of the few legitimate successors of Albrecht Dürer," the famous German Renaissance painter and printmaker (1471–1528). Kirchner moved to Davos, but his paintings were mostly exhibited and sold in Germany. Everything changed with the Nazi takeover in 1933. They decided who was and who was not a proper German painter. Over 630 Kirchners were removed from German art galleries. Some of them were shown in the special exhibition "Entartete Kunst" ("degenerate art"), which traveled throughout the country and attracted masses of people who would not usually go to museums. It was the perfect way to pillory someone. On June 15,

1938, Kirchner destroyed some of his paintings and sculptures and shot himself through the heart. He died with his German diploma of architectural studies in his breast pocket.

When trying to understand Kirchner's suicide, the traditional medical model would claim that Kirchner's long history of barbiturate and opiate dependence and the diagnosis of depression are the reasons for his self-chosen death. However, the point here is that we want to look beyond psychiatric diagnosis. The Nazis had destroyed *Kirchner's identity as the great renewer of German paining.* They had destroyed an existential life goal. And I am quite sure that the diploma of his early architectural studies in Germany that he carried on himself on the day of his death had a meaning. It was, at least, the other—official—German professional identity he had. Let me make this clear: Kirchner clearly had medical risk factors for suicide, but, as I have argued in the previous chapters, *it is the person who acts on suicidal thoughts and impulses, not the depression or the substance use.*

We learned about the meaning of life goals and personal needs in the previous chapter. Striving to become and sustain the person we want to be is a goal-oriented process. When I ask patients my standard question about what the most important thing in their lives is, I get answers like: "It is important for me to be a reliable and useful person, to help others, to be a good parent and provide security to my family, to prove that I can achieve something in life, that I can live up to my potential, that I am a useful person in our society, that I am loved and respected, that I am a self-contained person, to find happiness in life, etc." Life goals, as we know by now, are largely determined by our biography. When—like the violin maker in the previous chapter—as a child someone has repeatedly been put down and told that they are a failure, then we can see that such a person will be driven by the goal to be good at something, prove the father wrong, and strengthen one's self-esteem. However, existential needs that compensate for the adverse experiences in childhood come with risks. They make us vulnerable when we experience serious setbacks in the pursuit of our identity goals. For Kirchner, being humiliated by the German Führer was the ultimate assault on his life goal to become the legitimate follower of Albrecht Dürer. In the words of Roy F. Baumeister, a situation that deeply threatens our existential needs leads to a "deconstructed self" and to suicide as an "escape from self."[1]

The more suicide narratives I heard, the more it became apparent to me that suicidal thoughts and impulses occur when some negative experience seriously threatens our sense of identity. Personal identity, according to Erik Erikson, can be defined as a "sense of coherence and wholeness" in who we are, as well as in the eyes of the others, even when circumstances change.[2]

Identity goals and needs have a strong cultural component. In China, a major threat to personal identity is humiliation and loss of face. I remember the pictures of a dramatic scene I saw in 2011 in a Chinese newspaper: The young bride with the name of Li, left by her husband-to-be on her day of wedding, had tried to jump out of the window on the seventh floor of an apartment building. The family members and neighbors were desperately trying to keep her from falling by holding on to her white wedding dress. In Asian countries, shame and humiliation have been described as precipitants of suicide.[3] In South Korea, the high suicide rates of young people have been related to high family expectations to achieve academic success, part of that society's Confucian cultural legacy.[4] India experiences an epidemic of suicides by farmers after extreme weather conditions leave the farmers unable to repay loans, making it impossible to sustain their families.[5] In 2020, more than ten thousand people in India's agricultural sector ended their lives by suicide.[6] In

Hungary, high suicide rates are said to have their roots in a fatalistic tradition and a greater social tolerance of suicide.[7]

Insight gained: Existential needs are related to important life goals and our sense of identity.

Most of us will know shame from personal experience. It is an unpleasant inner experience related to some situation where we did not live up to our personal values or the social values of the culture we live in. Shame occurs when we have blundered. It is something we prefer to keep secret. Shame may seriously undermine our sense of who we are and who we want to be: to be loved, to be respected, to be reliable, to be successful, to be a good mother or father, or a good son, and so on.

Young people are particularly prone to shame. Oliver (eighteen) is the son of a traditional Catholic family; he has a younger sister and caring parents. The father is an architect, the mother an accountant in a local company. Oliver is in his second year of an apprenticeship in a travel agency. The parents notice that he looks somewhat preoccupied. When asked, he denies any problems ("No, why do you ask? Nothing's wrong with me, everything's in control, school and everything"). The deadline for the submission of his final diploma thesis approaches. Mother offers to see the manuscript—maybe she could make some useful comments. The deadline is Monday. On Saturday afternoon Oliver says he is going to the travel agency to print the manuscript. An hour later, he comes back to fetch paper for the printer—which mother finds rather odd, considering that he is working in an office where you would expect plenty of such supplies. Oliver does not come home. Two days later he is found in the forest. He had hanged himself.

Months later the parents came to see me. They were devastated, helpless, and trying to understand. They had talked to Oliver's friends and teachers. Oliver had not attended classes for weeks. They searched his laptop and found that he had not even started his thesis. Oliver's parents had not noticed any changes in their son that, even in retrospect, could have been signs of depression. Mother said that she had always told their children that it was important to talk about any problems, that they could tell her everything, and that lying was not the way to deal with problems, a parental attitude probably most of us would endorse. Most young men when they approach adulthood have a strong need to be self-reliant. Obviously, Oliver tried to keep up the image of a "normal boy" vis-à-vis his family. Yet, he had developed a construct of lies, which, the longer it lasted, the more difficult it became to overcome and tell the parents the truth. I am sure that his parents would not have made a scene had he managed to tell them the truth. For instance, he could have said "I need to tell you something. I have a problem . . ." It would have been as easy as that. We think.

Men, young men in particular, are extremely skilled in hiding problems that occupy their minds, to avoid cracks in their growing sense of identity. They may feel inadequate when comparing themselves with their parents or with cool peers. Weakness is something to be ashamed of. They develop a rather convincing identity toward the outside, the peers, the parents. They keep failure and weakness to themselves. What they try to appear to the outside is not congruent with what they feel inside.

In 2017, Mette Rasmussen and her colleagues from Oslo reported on the results of in-depth interviews with relatives and friends of ten young men who had died by suicide.[8] One of them was Adam. Adam had no record of mental illness, and before his death he had not asked for help or support from family or friends. The people close to him had not noticed any signs of serious confusion or depression in the last weeks before his sudden and unexpected death. His parents, his siblings, his male friends, and his ex-girlfriend were shocked. "How could he do such a thing—to kill himself?" When questioned, the family and friends saw a connection between the young man's need to prove himself as adult man among other men. Schoolmates and friends noticed that their friend had avoided any signs of weakness. They said that weakness was not allowed. Adam's suicide appeared to be the result of a young man who was unable to deal with his—perceived—shortcomings.

The Rasmussen paper refers to the story of another young man who took his life the same day he was meant to leave for a trip with an older friend. The day before the suicide, he said to friends and family that he had packed his bags and was looking forward to the trip. In the evening, he put an end to his young life. There were no signs of any preparation for the trip in his room; no bags were packed. We do not know why this young man was unable to say that he felt unable to join his friend and go on the trip. The suicide appears as a way to avoid losing face. These young men are not losers. They are often excellent and successful at school and in their studies—which, paradoxically, may make it especially hard to admit personal shortcomings.

Chapter 12 will look in more detail into the problem of suicidal behavior in young people.

"Psychological pain is an unbearable, destructive emotional state that demands resolution."[9] When the essential needs or life goals of a person are severely threatened by a negative experience, the consequence is emotional stress and psychological pain.

A patient wrote to me: "You as a physician must know that there is physical pain and mental pain. The mental pain is much stronger."

Edwin S. Shneidman called it "psychache":

> Psychache is the hurt, anguish or ache that takes hold of the mind. It is intrinsically psychological—the pain of excessively felt shame, guilt, fear, anxiety, loneliness, angst, dread of growing old or of dying badly. When psychache occurs, its introspective reality is undeniable. Suicide happens when the psychache is deemed unbearable and death is actively sought to stop the unceasing flow of painful consciousness. Suicide is a tragic drama in the mind. What my research has taught me is that only a small minority of cases of excessive psychological pain results in suicide, but every case of suicide stems from excessive psychache.[10]

For Shneidman, psychache was energized by threatened essential needs of the individual. Often, people find it difficult to describe their experience of pain. Israel Orbach spoke of a perplexing intrasubjective state that is experienced as negative inner mental events such as a sense of abandonment, estrangement, disintegration, going crazy, loss of self, self-hostility, and the like. He argued that mental pain is different from depression and from nonspecific stress. Bolger analyzed peoples' self-narratives to get a grasp of mental pain from an experiential perspective.[11] The descriptions included "brokenness of the self," with a sense of woundedness, disconnection, loss of self, loss of control, and a sense of alarm. Feelings of emptiness, loss of meaning, and a lack of perspective were present in most cases with suicidal behavior. When these negative emotions turn into a generalized experience of panic, when pain is experienced not only as unbearable but also as irreversible, suicide may appear as the welcome shortcut to end this mental condition.

Insight gained: Shame and pain are the main drivers for suicide.

In my search for mental pain literature I came across a remarkable book, *The Evolution of Suicide*, originally written as a PhD thesis by C. A. Soper, an evolutionary psychologist and psychotherapist. His theory is based on two elements: first, the experience of pain as an ancient, evolutionary biological stimulus that forces escape action and, second, the development of the human brain, that is, the large cortex, which enables *Homo sapiens* to produce sophisticated, goal-directed scenarios for escaping the pain experience—a combination that can end in "self-extinction." In Soper's words: "Suicidality probably arose in the human species as a noxious by-product of two adaptations combined: pain, an animalian biological signal that demands action to escape it; and the emergence of the human intellect, which offers intentional self-killing as a means to obey that demand for escape."[12]

Let's see what this could mean for our understanding of suicidal behavior. Mental pain as an existential threat to our existence triggers the amygdala, our alarm center in the brain. The amygdala sets off an alarm response (the fight-flight response), an acute physiological stress reaction to a life-threatening situation. The problem for *Homo sapiens sapiens*, however, is that the effect of the amygdala on cortical, goal-directed escape behavior is largely one-directional, which makes it difficult to control the alarm response and the suicidal urge.

I wrote to C. A. Soper:

> I am impressed. Your theory opens up a new and interesting dimension in explaining an enigmatic human behavior. The content fits extremely well into my model of suicide. You may know

that in our treatment model (ASSIP; www.assip.ch) unbearable mental pain is a key element which gets people into dissociative states, where the emotional brain takes over, overriding the cortex-based rational functioning. In our treatment it is crucial that patients learn about this mechanism in order to develop their warning signs and safety strategies. So, let me congratulate you on your book, the content of which matches very well with my clinical practice.

In chapters 2 and 3 we looked at the precarious connection between the emotional brain and the cortex, at stress and stress regulation, the power of emotions, and dissociation. Although the idea of suicide as an evolutionary phenomenon is convincing, in my view it doesn't take full account of the psychological aspects of suicide. We still need the person and the biographical background in any suicide model to understand what hurts and why. In my clinical work with patients I am so often impressed by how early suicide as a topic appears in a person's biography ("I first thought of suicide at the age of seven"; "my mother always threatened to leave the family and drown herself in the nearby river").

Psychological pain may be experienced by the teenager who is rejected by their first love, the high school student who fails an exam, the middle-aged woman who finds that her husband is having an affair, the fifty-six-year-old office employee who after thirty years of reliable work for the firm gets fired by the new boss, and so on. Most of these serious crises are triggered by related interpersonal experiences. More than once have I heard from patients I saw after a suicide attempt statements like: "I felt treated like a piece of dirt, not like a human being. It hurt so much, it was unbearable. This was when I decided to take my mom's painkillers to put an end to this pain. I hoped I wouldn't wake up again. I just wanted to end it."

But *inter*personal experiences can turn into *intra*personal experiences. Psychological pain is closely connected to negative thoughts about ourselves, lingering in our autobiographical memory: "I'm worthless, I'm a piece of crap, it is all my fault, I don't deserve to live, I'm a failure, things will never get better, I'm a burden to everyone around me, everybody hates me." This takes us back to Baumeister's *cognitive deconstruction*, which includes such typical "unfavorable attributions about the self." We have seen that early painful and traumatic emotional experiences can leave a person vulnerable in adult life. Early painful experiences are stored in our synapses. Actual experiences of the same quality, such as rejection, failure, etc., will reactivate old pain memories. In a World Health Survey in twenty-one countries, Bruffaerts and others found that the risk of suicidal behavior correlates with "childhood adversities" before the age of eighteen.[13] The more adversities, the higher the risk. Sexual and physical abuse were the strongest risk factors for both the onset and persistence of suicidal behavior, especially during adolescence, and the increase of suicidal behavior was at least threefold. Childhood adversities include all kinds of psychological trauma. In the case of Ms. M. in chapter 4, we saw how the traumatic separation from her mother led to a lifelong separation sensitivity ("when I don't know for sure that we will meet again, I panic"). For Ms. M., an event that seems negligible to most of us had the dimension of an existential threat. The girl who is devastated and takes an overdose after her first boyfriend dumps her has a history of her parents breaking up when she was five. "I felt it was because of me, it was my fault, and for years I felt guilty when I saw how my mother suffered."

Insight gained: Early negative experiences increase the risk for future events triggering mental pain and suicide.

Can we control the brain when it is in the alarm mode? In 1996, Joseph LeDoux, a neuroscientist at New York University, wrote a landmark book with the title *The Emotional Brain: The Mysterious Underpinnings of Emotional Life*. Let me pick out a sentence I consider to be crucial in our context: "Interestingly, it is well known that the connections from the cortical areas to the amygdala are far weaker than the connections from the amygdala to the cortex."[14]

LeDoux introduces an interesting concept: The brain has a "low" and "high road" in processing information (figure 7.1). Let me try to summarize. When confronted with signs of danger, the brain responds with behavioral, physiological, and endocrine (adrenaline!) reactions. As we have seen earlier, it is the amygdala that pulls the alarm. Information about a threat reaches the amygdala via the thalamus, which tends to be rather crude in its perception. This system, which has been passed on to us from the evolution of our species, has the disadvantage that in moments of danger it functions without involving our gray cells, the cortex. The advantage, however, is that it is fast. LeDoux uses the example of someone walking in the desert. A dry, clattering sound occurs. The information goes straight via thalamus to the amygdala and makes us retreat from the rattlesnake—a reaction that bypasses the cortex. This is the low road. In other words, the low pathway gets us to react to a dangerous situation before we can consciously respond to the situation. In contrast, the high road involves the cortical centers, where we consciously try to

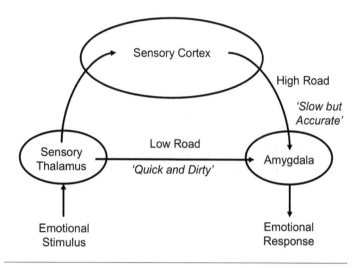

FIGURE 7.1. Low Road, High Road Model for emotional processing

Source: Adapted from J. LeDoux, *The Emotional Brain: The Mysterious Underpinnings of Emotional Life* (New York: Simon & Schuster, 1996), by Joost Heutink.

make sense of the situation. The high road (the sensory cortex) is slower and takes more time. Hopefully, the cortex will offer a more detailed and accurate interpretation of the situation and, if necessary, get us to override or modulate the rapid emotional reaction.

Modern *Homo sapiens* is not so much threatened by external rattlesnakes as by *internal* wild animals. Our psychological body is just as vulnerable—or even more so—as the physical body. We are, psychologically speaking, extremely vulnerable animals. The alarm system may be triggered by situations that threaten the sense of who we are. Unfortunately, we often find it extremely difficult to understand the language of our amygdala, particularly when there is no obvious enemy and when a negative experience links up with a vulnerability we acquired in childhood.

Yet we are not as helpless as it may appear. In chapter 4 we saw how Kevin Ochsner and his colleagues in their elegant study demonstrated that people can alter their responses to distressing pictures and how this can change the levels of activation in the amygdala. Treatment concepts developed for DBT (skills training) and for CBT (identifying negative automatic thoughts) are making use of these very same techniques for emotion regulation. MBCT, mindfulness-based cognitive therapy, teaches the suicidal person to apply techniques of stress reduction derived from meditation. If we take LeDoux's model of the low and the high road, it means is that when we are faced with a threatening psychological pain, we must learn to engage the cortex before we jump from the bridge. When suicide has been a topic in the past, we need to develop an early warning system that can effectively intervene before the emotion-driven autopilot program takes over. That is, our rational brain needs to learn to control the low-road emotional response at an early stage of the acute suicidal crisis and stop us from doing something we will most probably regret. Baumeister and Heatherton made a clear statement: "One of the most important human traits is the capacity of human beings to alter their own responses and thus remove them from the direct effects of immediate, situational stimuli."[15]

In the search for understanding the suicidal mind, we have now got a number of elements that belong together like the pieces of a slowly growing puzzle.

The elements are:

- Early negative experiences may result in personal vulnerabilities and existential personal needs.

- Later negative experiences of the same quality may trigger emotional stress and mental pain.
- High emotional stress may severely disturb our brain function, with the emotional brain in the alarm mode overriding the rational brain, which leads to the suicidal mode and symptoms of dissociation.
- The immediate goal of the suicidal person is to end and escape from this unbearable mental condition.
- People who are at risk of suicide need to gain insight into their personal suicidality; they need to know their personal suicide triggers, the warning signs, and how they can be safe in future suicidal crises.

Mr. K., the fifty-six-year-old engineer I interviewed after a serious suicide attempt (see chapter 3) said:

> In the evening I decided that this is enough. Tomorrow I'll end my life. I had dinner with my family, then we watched TV, I went to bed and slept quite well, we had breakfast together, then I took the car and drove to a remote place in the forest and swallowed all the pills I had prepared. It was winter, minus ten degrees, and then I fell asleep. Since from the evening it was like a set program, like functioning in an autopilot mode.
>
> This was not me doing that kind of thing. This is real frightening—I never want to get into this state of mind again.

One year later, when he had developed suicidal thoughts again, he drove directly to the psychiatric hospital he knew from his previous admission and said, "You have to take me in, I am suicidal." Twenty-five years later he is alive and doing well.

SUMMARY

In this chapter I have argued that a suicidal crisis occurs when our sense of identity is at stake. After an adverse experience we may be overwhelmed by extreme emotional stress, shame, and mental pain. In an evolutionary context, severe pain is associated with the impulse to escape. The emotional brain takes over and seeks a quick solution to this—seemingly—unbearable, existentially threatening experience. We have seen that in the alarm mode, rational thinking is shut down.

The good news is that we can teach our brain to control emotional impulses. To get there, patient and therapist must first develop a common ground, that is, a shared understanding of a person's drivers for suicide.

8

PERSONAL VULNERABILITIES AND SUICIDE

We have previously met the thirty-eight-year-old Ms. M., a mother of two children aged fifteen and thirteen, in chapter 3, where we looked at the EDA curve recorded during her interview. On admission to the medical emergency unit Ms. M. had been diagnosed with fractures of the pelvis, the lumbar vertebrae, ribs, the ankle, and a cerebral concussion. She had jumped from the balcony of her sister's third-floor apartment. After extensive orthopedic surgery she was seen by the duty psychiatrist on the surgical ward, and a diagnosis of a depressive episode with psychotic symptoms was made. She was put on antidepressant medication. Three weeks later, she was wheeled to our psychiatric department for the interview. On the Beck Depression Inventory at that time she scored 16 points, which falls into the range of a mild depression.

I began the interview with the standard opening question we used to start a narrative interview: "Can you please tell me in your own words how you came to the point of harming yourself? What is the story behind it?" She first said she couldn't remember much because she had been in a "strange mood." She described an inner feeling of arousal for several days, which began after her boyfriend had told her that he needed some time for himself, that is, it seemed to her that he wanted to take a break in their relationship. As a reaction she developed a strange urge to drown herself ("go into the water") in the nearby lake, an urge she managed to resist. The next morning she left home together with her son. She was driving, while her son was riding the bicycle. He took an unexpected shortcut, and she lost sight of him and panicked.

In the interview she started crying and said that separation had always been an issue throughout her life and that her children meant everything to her. She said that now, as she talked about it, she remembered that because of her urge to go to the lake she felt that something terrible might happen and her son might not see her again because she might not survive the day. The day before, she had tried to help her daughter with math homework but was unable to concentrate, and both daughter and mother ended up crying.

After she had lost sight of her son, she went to her sister's, who lived nearby, hoping that she could calm down. Her thoughts were racing and she felt she was about to lose control. She remembered trying to regain control by forcing herself to concentrate on eating a yogurt. Her sister later told her that all of a sudden she got up, "like a zombie," opened the door to the terrace and jumped. When she regained consciousness in the intensive care unit, she had a complete blank for nearly a week.

In the interview she continued talking about what separation meant to her. "This has followed me throughout my life—the problem with losing people whom I love. It doesn't matter who it is. I often had breakdowns with heartaches and anxiety attacks. I have great problems coping with separation and loss." She added that she did not know the meaning of saying goodbye, as she never knew if it would be for hours or forever. "My children mean a lot to me and are very close to me, and when I don't know for sure that we will meet again, I panic. And yes, my boyfriend said that he needed time, and I wasn't sure what it meant—if this meant total separation, never meeting again, or what. . . . This was how it started, and I got totally overwhelmed by this, although now he says he just needed time for himself. But for me time means. . . ."

PSY (INTERVIEWER): . . . something different?
PATIENT: It could have been days, years, or never again, as if he had said I don't want to have anything to do with you anymore. Since then there was this nervousness in me.
PSY: Your son had said that he would accompany you?
PAT: No, he had the same way to school, so we went together. He took the bicycle, and I was in the car. He then took a shortcut, which I wasn't aware of. And this was absolutely awful for me, because I didn't know if he would come back and whether I would see him again, as I hadn't said goodbye to him. Why I had these thoughts I don't know; I simply wanted to see him once more [cries]. . . . Already in the morning I had found it difficult to say goodbye to my daughter. Later, there were so many thoughts and they had such a power over me that I developed the feeling that I would really go mad. I then said to myself that I didn't want my children to end up with a disturbed mother or that they would have to come to see me in a psychiatric hospital but that they should rather have no mother at all, then. This was very strong, as if ending up dead or unconscious or I don't know what. I also thought that I would tell my sister that if I ended up as a vegetable they should switch off the machine, because I didn't want that my children or my relatives would have to suffer because I was nuts. This, too, I felt very strongly. Actually, I wanted to escape from these thoughts, not from the too heavy demands but from the many thoughts

because they made with me what they wanted. I couldn't live with them any longer. I wanted to kind of kill them.

We recognize the behavior before the suicide attempt as a typical example of dissociation. She was in a state of high arousal, no longer in control of her thoughts, feeling that she was losing control, unable to calm down, and, when she jumped, "acting like a zombie." Dissociation describes a special state of mind in which thinking, feeling, and acting are "dissociated," or disconnected. She couldn't remember the moment she got up or how she walked straight to the balcony and jumped.

Let us try to introduce some structure into Ms. M's. story.

1. A triggering event, which is a threat to a person's basic needs and life goals
    First, the boyfriend saying he needed time for himself
    Second, losing sight of her son on the way to school
2. The suicide-related emotions, thoughts, and actions
    A high level of arousal and emotional pain
    Symptoms of dissociation
    Thoughts of losing control
    The sudden behavioral impulse to escape from this unbearable mental state
3. Personal goals and basic needs, with their associated vulnerability
    The existential life goal to be a good and reliable mother
    The need for secure relationships
    The fear of abandonment (boyfriend, son)

The triggering event looks harmless to an outsider. For Ms. M. it was an intensely disturbing, life-endangering existential threat. What is missing here are the roots of this extreme fear of separation, which has a traumatic dimension (dissociation). We learned this later, during our second appointment—as we will see later in this chapter.

The fifty-eight-year-old Mr. T. was another patient I saw for one of the early narrative interviews. Mr. T. was admitted to the surgical ward of the University General Hospital after he had chopped off his right hand with an axe. In an eight-hour operation the surgeons had stitched the hand back to where it belonged—after it had been flown in by helicopter from Mr. T's remote home in the mountains, because the paramedics had forgotten to take it with them in the ambulance. Mr. T. told me the following story.

He chose early retirement from his work at the local post office and had been looking forward to a new life in small house he owned in a rural area. The first months went quite well; he was often out in nature, regularly visiting his eighty-six-year-old mother who lived nearby. He then started to notice slowly developing changes in his well-being: problems with sleep and appetite, withdrawal from others, impotence—"You must know that I have girlfriend, younger than me"—lack of concentration, and problems with memory. His thoughts started to go in circles. He started to think that his friends and colleagues would turn away from him, that they would realize that he was not the person they thought he was, that, in fact, he was an imposter and a crook. That he had always lived at the expense of others.

His girlfriend made an appointment for him at the local medical surgery. The family doctor was away, and Mr. T. was seen by a substitute doctor who took blood for a routine lab and gave him an appointment for the following week, when his regular doctor would be back. Two hours before this appointment, his condition worsened dramatically: He got into a state of panic, thinking that "the truth would now come out." He felt that the only solution was to "end it all." It was as if an autopilot program had taken over. He got a knife and cut deeply into his wrist. He was bleeding heavily but realized that this wouldn't work, so he got an axe and without thinking twice chopped off his hand. "I really wanted to go, this is all I was thinking." He was lying on the floor, waiting for his death. Then the phone rang. He first said to himself that he wouldn't answer, but after some time he crept to the phone. It was his girlfriend. She immediately called an ambulance.

In a usual interview, we might stop here because the story is coherent (all of what you just read took five minutes for Mr. T to relate). A man, after retiring, became isolated and developed typical symptoms of depression, including the typical depressive rumination on the deficits of his personality. Depressed people often fall into a negative cognitive spiral, as if the reservoir of negative and shameful experiences that had accumulated throughout their lives had burst a dam and flooded the brain. These cognitions subside when patients recover from depression. Any typical mainstream psychiatrist would see the depression as the cause of this life-threatening suicidal crisis. Yet I felt there was something missing in this story. Remember: Mr. T. got into the acute and dramatic suicidal crisis immediately before his appointment with his regular doctor. It didn't make sense to me.

PSY: This is a heavy story. Let me ask a few questions so that I can better understand all this.
PAT: OK.
PSY: So you didn't go to see the doctor. All this happened shortly before the appointment?

PAT: It happened shortly before. I had an appointment in the morning.

PSY: And you didn't go.

PAT: I was much too afraid. It would have been the second appointment. The first time it was just to take blood. My doctor was away on holidays the week before, and there was another doctor. This time I just got frightened. I was afraid that he would tell me what he found out from the lab results. It is the problems with memory that worry me. The results might have brought up something awful. I thought that if I went there, something would blow up.

PSY: Let me ask a bit more specifically. What do you think could have blown up?

PAT: That I had Alzheimer's.

PSY: I see.

[*Silence.*]

PAT: You must know that in my family an aunt of mine developed early Alzheimer's and ended up in a psychiatric institution. I was simply frightened that something like this might be discovered.

[*Silence.*]

PAT.: Then I . . . My girlfriend goes to the same doctor. I know there is this thing about confidentiality, but still . . . I worried that something would come out into the open. I was afraid that this would set something off. That in case he didn't tell me the truth he would tell her.

Here I need to add important information regarding the diagnosis of depression. The fear of dementia is not at all rare in depressed people, because, indeed, many of the symptoms of a severe depression can be similar to the signs of dementia (problems with memory and concentration, difficulty remembering names, etc.). The term used in psychiatry is "pseudodementia"—a condition that looks like dementia but in fact is a severe depression, which responds to antidepressant treatment.

I took him back to the suicide action.

PSY: After you had cut yourself—didn't it hurt?
PAT: There was some pain. But not much.
PSY: Then it was bleeding . . .
PAT: It wasn't bleeding enough. So I got the axe. I thought that then the blood would come faster.
PSY: You wanted to do it radically.
PAT: I got into a panic. I didn't think anymore. I simply thought "off and away."

I believe that cutting off one's hand with an axe is only possible in a state of dissociation. This is what Israel Orbach wrote about (see the box between chapters 5 and 6).

I asked Mr. T. to tell me about his life. He had left high school early and taken a job at the post office. He eventually rose to become head of a small local post office but felt he could barely cope with the job. He then changed to working at the till, where he had less responsibility but lots of social contact. He was married and divorced twice and had one daughter from each marriage. Both his wives left him. Both divorces had been extremely painful for him. When the second wife was about to leave him, he drove around aimlessly for hours, determined to kill himself in a car crash. The police found him on a forest road and took him to a hospital. For the past ten years he had now lived with his present girlfriend and her daughter.

We can now see that Mr. T.'s vulnerability—one side of the coin—is a deep-rooted problem with self-esteem. What I realize now, when writing about Mr. T. (who was one of my early cases with the narrative approach) is that the childhood roots of his vulnerability are missing in his story. What we can see, however, is other side of the coin: a need to be acknowledged by others as a valuable person. The breakdowns of his two marriages were a serious blow to this existential need. On the other hand, the social contacts with customers at the post office had helped him keep up his self-esteem—a personal validation he lost after retirement. The thought of being "found out" was unbearable to him. It meant that others would see the "truth": that he was a useless braggart (his own words).

The important elements in Mr. T.'s story are:

1. The triggering experiences
    Retirement and loss of social validation
    Increasing cognitive problems (severe depression)
    The appointment with his doctor
    The emotional stress caused by the fear "to be found out"
2. The related emotions, thoughts, and actions
    Negative self-evaluation; "it will come out into the open"
    Feelings of shame and pain
    The impulse to end this unbearable mental state
    A state of dissociation (autopilot mode)
3. The personal goals and basic needs, with their associated vulnerability
    The deep-rooted need to be acknowledged and valued by others
    The fear of the deficits of his self being uncovered
    The fear of being diagnosed with Alzheimer's disease
    The fear of being abandoned

Insight gained: Suicide may become a solution to experiences threatening our existential personal needs.

I need to add an interesting detail. Mr. T. was later admitted to the main psychiatric hospital in our city and transferred from there to a rehabilitation clinic. In a teaching round I showed the first five minutes of the video-recorded narrative interview. The psychiatrists from both institutions were astounded and said that they had not succeeded in getting Mr. T. to talk about his suicidal crisis. They could hardly believe what they saw in the video. "He seems to be another person." And they said another amazing thing: "We could never listen as long as you did in this interview, without saying a word."

The length of my "attentive silence" was exactly five minutes and nine seconds.

There is a personal inner logic that an interviewer and engaged listener is usually able to follow. The narrative approach, which I introduced in chapter 5, states that telling a coherent story to an attentive listener requires reflecting about oneself and others. This is the essence of "mentalization," a psychological concept that has been developed from attachment theory and the concept of the secure base (see chapter 6).[1] For interested therapists, I recommend the book *Mentalizing in Clinical Practice*, by Jon Allen, Peter Fonagy, and Anthony Bateman: "Mentalizing pertains to self and others: mentalizing is part and parcel of self-awareness and is thus essential to self-regulation, including coping with strong emotions; mentalizing in relation to others is crucial for healthy interpersonal relationships, and especially for developing and maintaining secure attachment relationships."[2]

Very early in our development of narrative interviewing, Ladislav came up with yet another idea—which, over time, proved to be an extremely powerful and novel means of joint mentalizing: The video playback and self-confrontation of patients with their narratives. Ladislav had gained experience in this from his work with Richard Young in Vancouver, where they had used video playback in health counseling with families. The replay of the video-recorded sessions activated the clients and got them to reflect and add more information about their experiences and problems. Some therapists had used video playback with their patients as early as the 1960s, but the technique was never established in clinical practice.

In the video playback, we jointly, therapist and patient, watched the video-recorded narrative interview. We first wanted to hear what patients felt about the narrative-style interview, so we decided that the video playback should be done by the one of us who had not interviewed the patient. We video-recorded all the video-playback sessions, so we could later analyze the content. The patients' evaluation of the narrative interviewing technique was—as expected—very positive. We then decided to use the video playback as an additional element in the working alliance between patient and therapist. We stopped the video every two to five minutes and invited the patients to comment on any thoughts, feelings, and sensations they'd had while watching the video. Very soon it became clear to me that this was an extremely powerful way of creating a truly active collaboration between patient and therapist. Most patients were glued to the screen, intensely following their own narrative—now in the role of the outsider, in contrast to the first session, where patients usually are reimmersed in their recent suicidal crisis.

Here is the transcript from the video-playback session with Ms. M.

PAT: Now as I see it again, it is as if someone else had told this story. I am less involved now because the story is told already, isn't it? I talked about it the way it really was and what I had felt at that time. In fact, it is my story.

PSY II: How was it for you to talk with the doctor?

PAT: Two insights. I realized how immensely preoccupied I was with the separation and how painful it had been for me. Separation is a very strong issue for me. Whenever I have to say goodbye to someone or whenever someone goes away from me. This has come out really strong.

PSY II: And this was about your son?

PAT: I had a need to say goodbye to him, as if I was never going to see him again.

PSY II: And when you talked about it you became tearful.

PAT: Yes. In a way I am surprised by this. You see, before it happened I felt that I didn't have any emotions in me anymore. I wasn't able to be angry or enthusiastic about something. I wasn't able to laugh or to cry. I had only been able to cry when my daughter had told me that she wasn't able to follow school any more and that I should really understand that the music lessons and sports was too much for her and that she couldn't cope with the situation any more. And these were the same emotions that came up in the interview. When talking about these two things, separation—and I feel that this is deep in me, my voice just now gets sort of blocked—and the inability of my daughter to cope, and that I hadn't recognized this. It makes me feel sad to talk about separation and the crisis of my daughter.

[*Ms M. cries.*]

I have the same feelings again, now. It is the same issue like on the video, that still touches me, saying goodbye.

PSY: It is the children that are important?

PAT: Yes, but not only the children. This has followed me throughout my life, saying goodbye to people I love. It is possible that this comes from my childhood, when mother left. Sometimes I say to the children that they must say goodbye like we would never see each other again, because we never know if we are going to see each other again.

PSY: How was this with your mother?

PAT: My mother left when I was seven. She left the family. She said that she had to go to a spa for her health, but she never came back. In fact she left the family for another man. I don't know if all this is connected. Then I knew that she would never come back, although she said she only left to go for a cure. And now there was this boyfriend of mine. He said that he needed time.

PSY: And this had preceded your crisis?

PAT: Yes, this was before, and I didn't know what it meant, if this meant temporary separation or final, or whatever. This was before and I was very much preoccupied by it, although now he says that he only needed some time for himself. But for me time . . .

PSY: Meant something different?

PAT: It could have been years or never again or no time for me anymore or wanting nothing to do with me anymore.

So, finally, in the video playback we find the missing piece of the puzzle. This is a perfect example of joint mentalizing, that is, the therapist becoming the coauthor of the patient's story. With the new knowledge of the severe separation trauma in her childhood, we can now easily see the inner logic of her life-threatening crisis. We understand now the meaning of "I have great problems in coping with separation and loss." Saying "goodbye" can mean anything from separation for hours or for forever. The boyfriend announced that he needed a "pause" in their relationship. This activated the "loss trauma," which became apparent in full force when losing sight of her son. To be a good and reliable mother to her children had become an existential need for Ms. M.: "When I don't know for sure that we meet again, I panic." Losing sight of her boy on the way to school acted like an "energizer" in a person who was already in a state of extreme emotional turmoil. When she realized that she was losing control over her emotions, the thought "rather no mother than a disturbed mother in a psychiatric hospital" appeared.

In trying to conceptualize the patients' stories, it soon became clear to us that suicide triggers and vulnerability are related to each other like a lock and a key. This is obvious in Ms. M.'s case, where we find an oversensitivity toward experiences of threatened loss and separation and the traumatic abandonment by her mother; in Mr. T.'s case, a deeply insecure person, it is the fear of being "found out" and rejected and the cognitive problems related to depression; and in the case of the George, the violin maker, it is the early experience of rejection by his father and the experience of being turned down by the violin makers' society.

Insight gained: Every suicide story is unique. What may be an existential threat to one person may meaningless to another person.

Because we wanted to find out what patients felt about the narrative interviewing style, we asked patients who were referred after attempted suicide to participate in a video-recorded research interview. They were told that the aim of the interview was to understand their suicidal crisis and that the interview would last about thirty minutes. The interview was followed by a short debriefing to ensure the patients' emotional well-being. Altogether, forty patients consented to this procedure and were allocated to two groups: twenty patients went through a usual clinical interview with one of my senior psychiatrist colleagues; the other twenty patients were interviewed by Ladislav, me, and Kathrin, a trainee psychiatrist who was familiar with our approach.

Right after the interviews, patients completed the Helping Alliance Questionnaire (HAq), an instrument to measure patient satisfaction with the therapist.[3] The eleven questions include statements such as: *I believe that my therapist is helping me, I have obtained some new understanding, I feel the therapist understands me, I feel I am working together with the therapist in a joint effort, I believe we have similar ideas about the nature of my problems.* Patients indicated on a scale from 1 to 6 their level of agreement. We included electrodermal activity (EDA, see chapter 3) as a measure to identify important life issues. Because of the time-consuming procedure of measuring EDA, we had to limit the final study report to eighteen patients. The study outcome was as follows: First, the interviews were rated significantly higher when in the opening sentence the interviewer used the words "story" or "tell me." The second finding was that interviews were rated significantly higher when the interviewer acknowledged the patients' important life goals and needs (as identified in the EDA), for example, "You know, separation has always been a big problem in your life." We concluded:

> A therapeutic alliance requires the active participation of the patient at every stage of a clinical interview, particularly so in the initial interview. The association of a narrative opening of the interview with HAq and relationship satisfaction is not surprising. Life-career aspects related to attempted suicide can best be accessed with a narrative approach, and, probably more important, the narrative account has in itself a therapeutic potential, giving the teller of the story the opportunity to give meaning to

experiences. The categorization was based on a simple concept: A narrative approach should manifest itself within the first few minutes by the interviewer inviting and encouraging the patient to tell the story with his or her own words.[4]

In our narrative study we had a senior colleague who was known to be very empathic in his professional attitude toward patients. In the video-recorded interview, the patient, a forty-year-old woman, gave him the lowest score possible. This didn't make sense to us, so we carefully watched the video. There was a sequence in which the patient says: "You know, I have been searching for love throughout my whole life." Clearly an important life issue, expressing her deep needs, and a very intimate statement at the same time. The therapist answers: "Ok, this is your perception. But—tell me now what happened the day before you tried to kill yourself?" Giving this empathic doctor low scores was obviously the response to his missing the importance of her mentioning her personal needs—and the associated vulnerability.

I learned three things from this study:

1. Listening and believing in the patients' narrative competence (the ability to tell a coherent story) is the best way to access a person's suicidal mind.
2. A narrative interview creates a strong therapeutic bond between patient and therapist, even in one or two sessions. It puts patients into an active role right from the beginning.
3. Any therapeutic encounter should be goal oriented. A first goal is to find a shared understanding in the context of a person's life story. The second goal will then be to collaboratively work toward changes that allow patients to better cope with future suicidal crises.

I have always been intrigued by the basic question of how on earth, with a "talking therapy," we can actually get patients/clients to change. Early in my training I probably read every article written by psychotherapy researchers such as Strupp, Luborsky, Horvath, Martin, Lester, Crits-Christoph, and others. I learned that the key to successful therapy is in creating a strong therapeutic relationship, or alliance, between therapist and patient. The following quotation nicely sums this research up:

> A major task for therapists is to form a working alliance with their clients so that they can go about the business of therapy as a joint, collaborative venture. Only by forming this kind of explicit therapeutic partnership is it possible to increase the likelihood that counselor and client activity during therapy will "mesh."[5]

The joint understanding of the patient's inner experience is essential. However, a meaningful therapeutic encounter needs more than that. It needs a shared therapy goal. It may well be that at the beginning of a therapy the goal and the motivation to working toward this goal are not yet there but that the change-oriented goals will emerge in the course of the therapy—as in my experience it is often the case with suicidal patients. In narrative interviewing, the rule is that an interviewer must never ask "why." This question presumes there is a simple answer and not a very personal story behind it. As outsiders, we tend to attribute causal explanations to a person's behavior and, in our case, to suicidal behavior. The patients, however, if we are open to listening, tell us stories of their inner experience of hurt, shame, self-hate, hopelessness, and unbearable pain. This issue has been addressed in the reason-cause discussion (see box 8.1).

Box 8.1: Reason-Cause Discussion

Buss (1978) argued that we have to distinguish the terms *reasons* and *causes* when attributing an explanation to another person's behavior.[6] Let's take an example. An eighteen-year-old woman has taken an overdose of painkillers and is admitted to the emergency room. She is tearful and says that her boyfriend has left her. For the ED staff it is clear: She took an overdose because her boyfriend has left her. Very obvious, isn't it? A causal attribution by an *outside* observer. The duty psychiatrist is called to determine how to proceed with this patient. The psychiatrist takes a seat next to the bed and opens the interview with "I am trying to understand. Can you tell me what is behind this that you wanted to put an end to your life?" Most probably, the patient will then give the psychiatrist an *inside* view of the matter. She may say: "I feel treated like a piece of trash that you throw away when you are fed up with it. I felt devastated, it must be my fault, there is something wrong with me, I am just not a loveable person. After all, this happened before. It hurt so much, and I had this awful pain I couldn't bear, my heart was pounding, I couldn't get enough air. You know it reminded me of the time when my parents split up when I was five. I thought it was because of me, and I felt guilty for many years when I saw how devastated my mother was."

Outside observers interpret the behavior from their own perspective, but they will not be able to understand the crucial individual context behind a person's suicidality without the active help of the patient.

At the European Symposium on Suicide and Suicidal Behaviour, in Bled, Slovenia, in 2000, I presented our model of accessing suicidal persons' stories in a plenary session. During the break a man grabbed my arm and said: "Thanks a lot for this presentation. It was the best talk in this symposium. You are talking like a social worker." I must have looked somewhat puzzled. He continued: "This is the biggest compliment I can make, because—you must know—I am a social worker." It took me some time to catch the significance of this compliment. But the meaning was clear: With our radically patient-oriented model, we had overcome the narrow medical model of suicide. Even though I was a medical doctor, I was talking in such a way that it reached patients as well as nonmedical health professionals. Indeed, what a compliment!

Insight gained: We need models of suicide that are easy to understand and meaningful to patients and therapists.

It was clear: It is not helpful for therapy with suicidal people to use models of suicidal behavior—and I am again referring to the medical model—that don't help people gain insight into their own personal suicidality. In a therapeutic discourse about the background of a suicidal action, the interviewer becomes the coauthor of the patient's narrative; that is, he or she is the "second author," helping to make sense of the suicide story, put some order into an experience of emotional stress, and gain personal insight.

The endpoint in the suicidal person's narrative is either suicide or life. When the story is told and retold to a sensitive listener, the endpoint will change from a death orientation to a life orientation: "Thus, in the case of the suicide attempter seen after admission, in a therapeutic interview, the narrative not only has a beginning, a middle and an end, but the prospect of an alternative solution."[7]

A narrative will open up possibilities to explore new coping strategies for future suicidal crises. Any death-oriented action step of the suicidal process can be replaced by alternative, life-oriented actions, thus indicating that no order of action steps must inevitably lead to a suicide. The suicide-as-action model is directly related to the issues of choices and decision making. Treatment following a suicide attempt will therefore have to help people regain a sense of agency and choice, as a form of empowerment. For a safe future, they will need to establish new strategies that will be readily available to escape the development toward suicide before the stress system shuts down the prefrontal cortex and the dangerous goal-directed suicide autopilot takes over. Let me quote Tali Sharot, a cognitive neuroscientist at University College London. In her 2017 book *The Influential Mind*, she wrote: "If we want to affect the behaviors and beliefs of the person in front of us, we need to first understand what goes on inside their head and go along with how their brain works."[8] This is what should guide professional helpers.

The narrative approach is not limited to suicidality. *Narrative medicine* is the application of the narrative approach in somatic medicine, promoting the narrative of the patient's illness experience and dealing with symptoms and problems of health: "A medicine practiced with narrative competence will more ably recognize patients and diseases, convey knowledge and regard, join humbly with colleagues, and accompany patients and their families through the ordeals of illness. These capacities will lead to more humane, more ethical, and perhaps more effective care."[9] I couldn't agree more. We need a revival of what used to be called "good clinical practice."

SUMMARY

In this chapter we have seen how crucial it is to listen to the patients' stories in order to collaboratively come to an understanding of each suicidal person's inner logic. We have seen how a suicide story is not complete without the related biographical background. Suicidal patients need to gain insight into their vulnerabilities and personal needs. A clear, person-centered approach is the basis for a strong therapeutic working alliance—the prerequisite for a therapy to be effective. In the next chapter I need to pay tribute to the participants of ten years of regular international meetings in Switzerland dedicated to exactly this: sharing and further developing person-centered therapeutic approaches to suicidal patients.

9

A THINK TANK OF CONCERNED THERAPISTS

In the year 2000, my dear colleague Ladislav and I invited a small handpicked group of health professionals to Switzerland to discuss our person-centered and action-based model of suicide. A study grant we had received from the Swiss National Foundation of Science covered the costs for travel and accommodation. The Aeschi conferences developed from this three-day meeting into a forum where new ideas and concepts were shared at a high level of professionalism. In a unique and intense working atmosphere, dedicated therapists and researchers, faculty members as well as participants, collaboratively contributed to developing and improving treatment models for suicidal patients. With these biennial international conferences, our small team was catapulted into another sphere.

Houston, April 15, 1999, the annual meeting of the American Association of Suicidology. The first plenary speaker was David A. Jobes, acting president of AAS. I had never heard of him. The title of his presentation was "Collaborating to Prevent Suicide: A Clinical-Research Perspective." He started his talk by highlighting the problems health professionals have in dealing with suicidal patients, not sparing his colleagues. Dr. Jobes said: "The majority of suicidal patients who seek mental health care are at some level seeking a relationship that will help bring them back to life instead of a death trajectory. However, what such patients may well encounter is a clinician who is unhelpful or even harmful because of unavailability, elusiveness, or authoritarian response to suicidality."[1]

It was a truly masterly and courageous talk. He showed a figure that still today I include in all my presentations. The figure juxtaposes the traditional medical interviewer and the empathic interviewer (figure 9.1). In the top picture the interviewer understands suicidality as a symptom of depression. The interviewer is the expert, sitting higher up; assessing psychiatric pathology; asking questions about sleep, appetite, loss of energy, worrying, etc.; and deciding on a diagnosis. In the lower picture, patient and interviewer are both seated side by side, closely collaborating toward a joint understanding of the patients' inner and very personal experience. Here, the conversation is about shame, pain, agitation, self-hate, hopelessness, etc. A new definition of patient and therapist roles. Dr. Jobes continued: "Central to the protocol is the emphasis on placing the clinician and patient on the same team, working together to understand how suicide has become so viable and compelling in the patient's life."

I wasn't sure what the audience made of this, but I was excited, and it was clear that I had to link up with this man. As soon as he had finished with his presentation, I went up to him and told him how excited I was hearing his ideas. I told him about our narrative interviews and the video playback—therapist and patient sitting side by side and collaboratively working toward a joint understanding of the patient's suicidality—exactly the illustration he had used in his presentation. Would he come to a

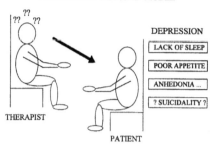

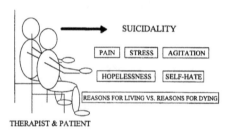

FIGURE 9.1. The medical vs. the collaborative model.

Source: D. A. Jobes, "Collaborating to Prevent Suicide: A Clinical-Research Perspective," *Suicide and Life-Threatening Behavior* 30, no. 1 (2000): 8–17.

small international expert meeting in Switzerland, together with some likeminded health professionals, to discuss our model of understanding suicidal behavior? Yes, he would love to come.

During the coffee break, I spoke to Lanny Berman, executive director and gray eminence of the American Association of Suicidology, whom I knew from earlier conferences, about my plans. He didn't think twice but grabbed my arm and introduced me to Dr. Maltsberger—"he is the man you want to meet!" John T. Maltsberger MD, a psychoanalyst from Boston, was an elderly man, interesting, highly educated, and captivating. His lively eyes would often start to glow when he was about to make a somewhat mischievous and provocative remark—something he loved to do. Yes, he agreed, he would love to come to Europe for a meeting. So did Antoon A. Leenaars, a Canadian psychologist of Dutch origin and past president of the AAS. He, too, had strong ideas about psychotherapy with suicidal people. He was a friend and collaborator of Ed Shneidman, the American icon of suicidology. Next, I invited Israel Orbach, an outstanding clinical psychologist from Israel, a dedicated psychotherapist with a deep interest in people, a real character, someone you could not overlook. He, too, knew Shneidman and had worked with him.

Bern, February 17, 2000, the Restaurant Hotel Schweizerhof. I was waiting for my guests to arrive. And, incredibly, they all turned up, first David Jobes and John (Terry) Maltsberger, then Israel Orbach and Antoon Leenaars, then my colleague Ladislav Valach with Richard A. Young from Canada. Richard had been working with Ladislav on action theory and health counseling. He was the one who had introduced the technique of video playback (or self-confrontation) into counseling. We took the train toward Interlaken and the Alps. I had picked a special place for this meeting, Hotel Aeschipark, a cozy hotel in a small village with chalets on the edge of the Swiss Alps. Little did I know at that time what role the hotel and the name Aeschi would play in the years to come. In the evening, my two young colleagues who had been on my team joined us: Kathrin Stadler, who had written her MD dissertation on our therapeutic alliance study, and Pascal Dey, a psychology student who had participated in our clinical studies.

The working conditions couldn't have been more unique. Overnight a blizzard arrived, and it continued to snow for two days. Imagine nine people gathering in the warmth of a cozy meeting room, the snow falling silently outside, engaged in watching our video-recorded interviews, commenting on the patients' stories, and confronting one another with different models of understanding suicidality. And, what was even more exciting, discussing the interviewers' therapeutic attitudes and interventions. We also showed the videos of the video playback and the patients' own evaluations of the interview, the Helping Alliance Questionnaire, which I introduced in the previous chapter.

The title I had chosen for this meeting was: "Understanding Attempted Suicide: Do We Need a New Approach?" The discussions were intense and incredibly stimulating. It was as if the small group of experts was electrified. The interpretations of the cases differed widely, from a psychoanalytical view, to the "Shneidman school," to cognitive theories of behavior, and to our model of suicide as a goal-directed action. One of Shneidman's dicta had been that suicidal people hide behind a mask and that the therapist has to "reach through the mask." With our approach to the suicidal patient, we had not encountered any masks to reach through. With our narrative interview approach—in which the patients and not the interviewer were the experts of their stories, most patients did not appear to have problems telling us, in a meaningful and understandable way, their suicide-attempt story. Ladislav maintained that, with an attentive listener, the patient's goal would be to give the interviewer "a good story," while the therapist's goal was to help the patient achieve this by being empathically and fully focused on the patient's story.

On the first day, John Maltsberger, the psychoanalyst, objected to our claim that a narrative interview could create a therapeutic relationship in a single session. On the second day, he said to me: "Well, after forty years of practice I think I'll go home and try your way to do interviews." Later, in a book chapter titled "Empathy and the Historical Context, or How We Learned to Listen to Patients," he wrote:

> The listener is empathic and promotes the patient's active participation in telling the story—the life experience is the patient's, after all—who knows it better? From the beginning the interviewer fosters the establishment of a therapeutic alliance, pays particular attention to the patient's subjective emotional experience, and at once contains and makes use of his countertransference experience. It is an approach not dominated by any particular theoretical perspective, and might be called, alternatively, the approach of intersubjective empathic attention.[2]

The day Dave Jobes arrived back home in Washington, DC, I received an email: "Dear Konrad, this was the best meeting I ever attended. There is important work in all this that we must get out into the field. Best regards, Dave."

The group soon agreed to continue the work as the Aeschi Working Group, a group of health professionals dedicated to promoting a truly patient-centered model of understanding suicidal behavior (figure 9.2). What followed was ten years of biennial Aeschi conferences in the same village, the same hotel, organized on my private budget with the help of a university fund and other sponsors. Aeschi soon developed into a magnet for therapeutically minded professionals, a meeting unlike any other. For the program it had once again been the idea of Ladislav to start every morning of the three conference days with video-recorded patient interviews—"after all, this

University Hospital of Social and Community Psychiatry Bern

The Aeschi Working Group
MEETING THE SUICIDAL PERSON
The therapeutic approach to the suicidal patient:
New perspectives for health professionals

2ND AESCHI CONFERENCE
UNDERSTANDING AND INTERVIEWING THE SUICIDAL PATIENT
6 - 9 MARCH 2002, HOTEL AESCHIPARK, AESCHI, SWITZERLAND

FIGURE 9.2. Aeschi header.
Source: From the author's archives.

conference is about patients and their stories, isn't it?"—shown in the plenary sessions. After fifteen or so minutes with the video, two discussants, usually with different therapeutic backgrounds—say, a cognitive behaviorist and psychoanalyst—presented their ad hoc interpretations of the case and the patient-interviewer interaction. The case discussions were then opened for the audience, clearly a highlight of Aeschi. The afternoons were filled with parallel workshops and poster presentations, followed by after-dinner presentations.

Unlike other congresses, all participants resided in the hotel, looked after by Ursula and Rainer Friedl and their staff, including the hotel cat, which would usually settle down at the legs of the presenters. Soon the small conference attracted mental health professionals from nearly every corner of the world, from most European countries, including Eastern European countries, the United States and Canada, Russia, India, China, Australia, Jamaica, and so on. In some of the conferences—which we limited to a maximum of seventy participants—we hosted health professionals from sixteen countries. Participants could simply not avoid being mesmerized by the "Aeschi spirit," an atmosphere full of caring, sharing, and lively discussions, which continued over dinner and later at the hotel bar, often leading to new collaborations. There was a family feeling, and it was no surprise that many participants, especially those from overseas, were regular attendees for years. The American colleagues were particularly open to present and put their concepts and new treatment approaches forward for discussion. Without them Aeschi would not have survived.

At the end of a conference, Savi Anthony from New Zealand presented a poem she had written the night before.

> Breathtakingly beautiful
> Amidst snow clad mountains
> Is the little village of Aeschi
> Unspoilt and pure
> Tucked away from the world
> Nations gathered
> People mingled
> Ideas shared
> Concerns discussed
> Thoughts developed
> To serve others
> In alleviating anguish
> Interactive and dynamic
> Knowledgeable and innovative
> Research, lectures and workshops
> Presented with clarity
> Inspiring us to move forwards
> Motivation maintained
> Enthusiasm retained
> Curiosity aroused
> Hope rekindled
> Hearts glowing warmly

We take our leave
Feeling happy and fulfilled

You wouldn't guess that the conference was all about suicide!

"Discovering the Truth in Attempted Suicide"—a title Dr. Maltsberger had insisted on—was a paper published in 2002 in *the American Journal of Psychotherapy*.[3] We had worked hard on formulating guidelines for clinicians (see chapter 13), which still today are a valuable resource, particularly in times where suicide risk assessment has developed more and more into a formalized cookbook interrogation to which patients can only answer with "yes" or "no." In one of the Aeschi conferences a participant got up and said: "This is wonderful work you do, but, to be honest, it is not really something new." I said "I agree. There is a name for it. It is called good clinical practice. It is about listening to our patients and developing a working alliance toward a shared goal." Later, conference participants called it the "Aeschi Philosophy."

When in 2012 I retired from the university, my American colleagues felt that Aeschi couldn't die. Thanks to Michael Bostwick and the Mayo Clinic, the biennial Aeschi conferences have continued since 2013 in Vail, Colorado. Partly because of the COVID-19 pandemic, the 2019 Aeschi conference appears to have been the last one. However, Aeschi continues to be present in many publications, for example:

> This understanding of the function of the suicidal wish appeared to facilitate genuine empathy for the client and connectedness in the relationship. This is in line with the fundamental approach

promoted by the Aeschi working group (Michel, 2011) and the CAMS (Jobes, 2016); which views the suicidal thoughts, sensations and behaviours as actions that contain personal meaning for the client.[4]

and

The Aeschi Working Group—founded in 2000—strongly emphasized the client-oriented approach, putting the client at the center of the stage as expert of his/her story, to tell about their own suicidality (Michel & Jobes, 2010). This approach also underlined the importance of the therapeutic alliance and understanding the client's conceptualization of suicidality. It originated in opposition to the traditional approach on risk assessment and treatment of the suicidal person, considered to be too medicalized and impersonal.[5]

I need to pay tribute to the many wonderful Aeschi contributors.

David Jobes, professor of psychology at the Catholic University of America, Washington, DC, was the great supporter and co-organizer of the Aeschi conferences. His mastery of empathic understanding of the suicidal patients in the case discussions was awesome, and his concept of the collaborative therapeutic relationship has deeply influenced me. It is beautifully expressed by the following sentence: "I want to see it through your eyes. The Collaborative Assessment and Management of Suicidality (CAMS) is a clinical framework that guides the identification, assessment, treatment, and follow-up of suicidal patients."[6]

Israel Orbach, professor of psychology at the Bar-Ilan University, Ramat-Gan, Israel, was a clinician with an impressive empathy for suicidal individuals. He was the one who taught me to not be afraid of patients talking about their suicidal wishes. He would tell his patients that they had to convince him that there was no other solution than their suicide. Truly genial. The teller can't stop putting themself into the listener's mind, which, particularly when the acute suicidal phase is over, means that people start again to see—like the interviewer—that there indeed *are* other possible solutions. In addition, talking to an empathic listener reduces stress, which defuses the suicidal mode and reactivates "normal" problem-solving capacities. Israel was one of the few experts who knew all about dissociation and altered body experience in self-destructive behavior. Israel died November 2010. Here is a typical Orbach statement:

> Being empathic with the suicidal wish means assuming the suicidal person's perspective and "seeing" how this person has reached a dead end without trying to interfere, stop, or correct the suicidal wishes. This means that the therapist attempts to empathize with the patient's pain experience to such a point that he/she can "see" why suicide is the only alternative available to the patient.... Instead of working against the suicidal stream and trying to instantly increase the patient's motivation to live by persuasion or commitment to a contract, the therapist takes an empathic stance with the suicidal wish and brings it to full focus.[7]

John T. Maltsberger, professor of psychiatry at Harvard Medical School, an accomplished and experienced psychoanalyst, brilliant in grasping the psychological dynamics of suicidal patients, was *the* senior Aeschi figure. His plenary case discussions were the highlights of every Aeschi conference. Few people knew so much about affective states preceding suicide, such as desperation, rage, anxiety, abandonment, self-hatred, anguish, unbearable pain, and dissociation. For quite some time he tried to dissuade me from inviting the cognitive behaviorists. But then, to my great surprise, he, the psychoanalyst, told me that he had booked a training course in DBT (dialectical behavior therapy) run by Marsha Linehan. An example of the effect of the Aeschi spirit. Terry died in October 2016. He was the founder of the Boston psychoanalytic group, a wonderful gang of deeply engaged psychotherapists who attended each Aeschi conference: Mark Goldblatt, Elsa Ronningstam, Mark Schechter, and Igor Weinberg, a clever young colleague who made his career at McLean Hospital, in Massachusetts.

J. Michael Bostwick is professor of psychiatry at the Mayo Clinic, Rochester, Minnesota, and the Mayo Foundation for Medical Education and Research. He has a psychoanalytic background, and he was one of the very early aficionados and regular attendees of Aeschi. His on-the-spot brilliant case discussions in the plenary sessions following the patient videos were legendary. From 2013, he continued the tradition of the Aeschi conference in the United States.

M. David Rudd, professor of psychology, a cognitive therapist trained at the Beck Institute in Philadelphia, impressed everybody with his cogent concepts of cognitive therapy. He was first author of the landmark cognitive therapy book for suicidal behavior.[8] It was at an Aeschi conference where for the first time I heard of concepts such as the *suicidal mode*, the cognitive belief system, and suicide triggers. As a rather autocratic conference organizer, I began to give the cognitive behaviorists more and more room because I felt that these therapy concepts were indeed promising.

Gregory K. Brown, associate professor of clinical psychology at the University of Pennsylvania, is another cognitive therapist and a dedicated Aeschi person. Gregory Brown is the main author of the first large randomized controlled study on cognitive behavior therapy (CBT) for the prevention of suicide attempts published in 2005.[9] At the eighteen-month follow-up, the treatment group had significantly fewer suicide reattempts and a 50 percent lower risk to attempt suicide. This study had a special meaning for me as a model of a top-quality academic achievement (something, which at that time, with my primary identity as clinician, I never imagined I would ever get close to). At the 2009 Aeschi conference Greg created a small turmoil. In a workshop he presented his concept of imagined reexposure of patients to the suicidal crisis as integral part of the cognitive behavioral therapy model. Some workshop participants felt that it was too risky if not outright unethical to reactivate the suicidal crisis and the suicidal mode. Little did I know then that ten years later I would use reexposure in our own concept of brief therapy. In cognitive behavior therapy, exposure is a central concept in learning a new behavior. I soon realized that there was no reason to be afraid of reactivating the suicidal crisis. In fact, avoiding the issue would be wrong—this is what people usually do. It is much better to face the bull in a safe environment.

Jeremy Holmes, a British consultant psychiatrist and psychotherapist affiliated with the Psychoanalysis Unit, University College London, was the unmatched expert on the patient-therapist relationship seen in the frame of Bowlby's attachment theory (see chapter 6). Attachment theory "regards the propensity to make intimate emotional bonds to particular individuals as a basic component of human nature, already present in germinal form in the neonate and continuing through adult life into old age."[10] Jeremy Holmes's work about the importance of the secure base in a therapeutic relationship is unsurpassed. "Therapists aim to create some of the parameters of secure base in their dealings with patients: consistency, reliability, responsiveness, non-possessive warmth, firm boundaries. This, it is hoped, becomes internalized as a 'place' in the psyche to which the patient can turn when troubled, even after therapy has come to an end."[11]

Bowlby wrote that a strong and effective therapeutic relationship "is analogous to that of a mother who provides her child with a secure base from which to explore the world." The therapist is reliable, attentive, responsive to the patient's narratives, and strives "to see the world through the patient's eyes" (note the similarity to Dave Jobes's statement earlier!). Feeling secure helps the patient put feelings into words. The therapist's task is to provide the patient with a "thinking mind." The goal is that the thinking mind function may eventually be internalized by the patient and foster the "self-reflexive capacity." This requires a therapist to be highly sensitive to the needs of the patient: "Like an attuned parent, the effective therapist will intuitively sense when the patient needs stimulus and direction to keep the thread of narrative alive, and when she needs to be left alone to explore her feelings without intrusion or control."[12]

Secure attachment is associated with narrative competence. Insecure and, above all, disorganized attachment characterizes borderline personality syndrome and is related to an incoherent narrative style, usually linked to early trauma or neglect. Considering my early imprinting with attachment theory during my training years in the United Kingdom, the attentive reader will understand why for me Jeremy Holmes was a very special member of the Aeschi faculty.

Jon G. Allen, PhD, professor of psychiatry at the Menninger Department of Psychiatry and Behavioral Sciences at the

Baylor College of Medicine, Houston, Texas, introduced further developments of attachment theory and presented the *mentalizing* concept, that is, the ability to reflect about mental states (e.g., emotions, intentions, and goals) of oneself and others (see chapter 8). The mentalizing capacity for people with problems of emotion regulation is state dependent; that is, when emotions and stress hormones shoot up, people lose their capacity to mentalize and to rationally analyze interpersonal conflicts. The practical aspects of the therapist's attitude to foster mentalizing are brilliantly explained in the book *Mentalizing in Clinical Practice*.[13] What has particularly found its way into my therapeutic arsenal (and which is crucial for narrative interviewing) is the interviewer's "not knowing" attitude, which stands in clear contrast to an interviewer who is "striving to be clever, brilliant, and insightful."

Marsha M. Linehan, professor of psychology and director of the Behavioral Research & Therapy Clinics (BRTC), University of Washington, Seattle, attended Aeschi twice and was an impressive authority on self-harm and emotion regulation. Marsha Linehan developed DBT—dialectical behavior therapy, which is widely known and practiced, a specific therapy program for individuals diagnosed with borderline personality disorder, a condition with a high long-term risk of self-harm and suicide and that is generally difficult to treat. Such individuals have a greater sensitivity to emotional stimuli and a greater intensity of emotional experiences. DBT was probably the first cognitive behavioral therapy that has been shown to be effective in reducing self-harming behavior. DBT focuses on factors contributing to suicidal behavior and aims to change the behavior in the present, developing individual treatment plans.[14] In DBT, therapists take an active and instructive role—yet another contribution to the range of therapy models presented at Aeschi.

Lisa Firestone, clinical psychologist and director of research and education at the Glendon Association, California, was a regular Aeschi attendee. She convincingly presented the concept of the negative inner voice and showed powerful videos of treatment sessions. I vividly remember an impressive case of a woman who in the acute suicidal crisis recalls being haunted by the negative inner voice of her aggressive father—another way of linking mental pain and suicidal behavior to early painful memories, that is, the person's biographical vulnerability. Negative thoughts about the self are often related to early traumatic experiences, expressing themselves as destructive inner negative voices attacking the self—a theory that appeared to be in line with Baumeister's blaming of the self and the urge to escape the threatening thoughts.

James R. Rogers, from the University of Akron, Ohio, added an important tile to the emerging mosaic of understanding and helping the suicidal person. He argued that a one-size-fits-all approach in the typical crisis intervention with suicidal individuals is not effective in preventing suicide but, much worse, could even contribute to suicidal behavior.[15] His point is that, after a crisis intervention following a suicide attempt, if clinician and patient both assume that the problem is now solved, both are wrong, and detrimentally so. The patient will not be aware that the risk will from now on be increased, and the therapist will not help the patient understand how to deal differently with future life-threatening crises. Rogers argued that the goal of therapy must be to see each suicide as a deeply human and unique experience. James Rogers died in 2013.

Richard A. Young, from the Department of Educational and Counselling Psychology, University of British Columbia, is a longstanding collaborator of Ladislav Valach. He originally came from outside the circle of suicide experts. He had developed the video-playback and self-confrontation technique (see chapter 8) in health counseling.[16] Ladislav had convinced me that video-based self-confrontation (video playback) was an enormously useful tool. It would later play an important role for the Bern therapy model.

Ladislav Valach, my dear colleague and mentor, I don't need to introduce here, as the reader knows what eminent role he played in developing the Aeschi approach and the Aeschi conferences. He is an incredibly creative mind and has continued to publish on the topic of action theory and suicide.[17]

When planning for the next Aeschi conference, I always made sure that suicide was represented as a complex human phenomenon for which there is no simple explanation. Besides the psychotherapy-oriented clinicians, I invited experts in brain research, genetics, psychopharmacology, and more. There were workshops on mindfulness-based cognitive therapy (MBCT), the Next Day Appointment, controversies about the usefulness of inpatient care, combat-related trauma and suicide, therapists as survivors of a patient suicide, and many more. I hope that the many Aeschi contributors I did not mention here can forgive me.

The icing on the Aeschi cake was the volume *Building a Therapeutic Alliance with the Suicidal Patient*, edited by Michel and Jobes.[18] The book is a summary of contributions representing the essence of the Aeschi spirit, from the psychoanalytic understanding of suicidality (Maltsberger, Goldblatt, Schechter, Weinberg, Ronnigstam) to general questions of psychotherapy for suicidal people (Leenaars), to action theory (Valach, Young), collaborative therapy and narrative interviewing (Jobes, Michel), attachment theory (Holmes), mentalization (Allen), cognitive behavior therapy (Brown, Rudd), DBT (Linehan, Rizvi), questions of hospitalization (Lineberry), and pharmacotherapy (Bostwick). Marsha Linehan wrote a perfect foreword. She personally knows how it feels to be trapped in a suicidal crisis.

In spite of beautiful patient-oriented psychological theories, I realized that we had to be careful not to lose sight of other factors relevant for treatment, above all, the psychiatric diagnoses. Although I do believe that the old retrospective studies, which reported that 93 to 95 percent of suicides fulfill the criteria of a psychiatric disorder, are basically still valid, the overall percentage has been increasingly questioned. Affective disorders, psychotic disorders, substance use, and personality disorders are without doubt frequent in people who die by suicide. Furthermore, as we have seen in chapter 4, suicidal behavior has (neuro) biological correlates, and neurobiological research has helped us understand how suicidality is related to brain function. For each suicidal patient it is important to make a proper diagnostic assessment and to initiate, if indicated, a proper pharmacological treatment.

I need to stress that the "Aeschi philosophy" does not claim to provide a new theoretical model of suicide and suicidal behavior. The Aeschi approach is a model to understand the single suicidal patient. It is the suicidal person who acts, and an effective therapy needs to be person centered. As David Jobes says: "I want to see it through your eyes." We need a meaningful working relationship, and patients need to be actively involved in treatment. If we understand suicide as an intimate and personal experience of the patient, then the only way to establish trust is to take a genuine interest in the person and to get this person to tell us the story behind the suicidal crisis. If Kandel (chapter 4) is right, then an effective therapeutic process creates new synapses. Simply stimulating serotonergic synapses in the PFC with antidepressants is not going to result in a long-term behavioral change.

SUMMARY

In this chapter I wanted to show that new concepts do not develop in isolation but are usually the result of collaboration among experts. This was the hallmark of the Aeschi conferences. I am still indebted to the many Aeschi participants who freely shared their experience and knowledge. Everybody contributed and helped create a unique atmosphere ("the Aeschi spirit") where we could learn from one another and go home with new ideas. We were lucky to have some of the world's leading clinical experts for the treatment of suicidal individuals as regular and active participants. The exciting exchange from the Aeschi conferences finally encouraged me and my colleagues to translate the collaboratively gained knowledge into a structured therapy program that should be brief and effective at the same time.

10

TRANSLATING ACQUIRED KNOWLEDGE INTO A NEW THERAPY

How do we design a novel therapy with what we have learned so far? Here are the conceptual ingredients we agreed on in our team:

- Suicide is an action.
- Actions are carried out by persons who have their reasons to do so.
- Suicide usually happens in an out-of-the-ordinary mental state.
- During the acute suicidal crisis, our brain function is radically altered.
- The emotionally stressed brain loses the ability think and decide rationally.
- Suicide is taking action—but taking the wrong action.

However,

- We can learn to become aware of impending danger.
- We can learn to change our response to emotional impulses.

I then discussed with my colleagues how this could be translated into a structured therapy program.

First, considering the limited resources of professionals in the treatment of suicidal patients throughout the world, it was clear that a new therapy would have to be brief and well structured. Second, the therapy program had to be based on, first, our person-centered approach to the suicidal patient, with the Aeschi guidelines as the centerpiece, and, second, the concept of understanding suicide as an action. Third, the therapeutic alliance clearly had to be a hallmark of our therapy program. I have always been convinced of the therapeutic power of a truly collaborative approach. Attachment theory had taught me that people have the potential to gain insight and to change when they feel understood and safe in a therapeutic relationship (the secure base experience, see chapter 6). In this model, the therapist's task is to provide the ground for the patient to develop new ways of responding to emotional crises. We had learned that in order to establish a working alliance, the therapist had to abandon the typical health professional's role of being the expert on all mental health matters—*because suicidality is not an illness*. Thanks to the enormous persistence of Ladislav, I had learned that in a meaningful interaction only the patient and nobody else can be the expert of their life story and, in particular, the suicide story.

Fourth, with our psychiatric department being close to the emergency department of the General University Hospital, it was obvious that a new treatment should be designed for patients who had attempted suicide—the population with the highest risk of suicide. The 2005 Brown et al. study served as a model—showing that with a structured and time-limited therapy program it was possible to reduce the risk of reattempts. The great challenge, however, would be to create a much shorter therapy program that would be just as effective. Regarding the treatment goals, we agreed that we could not "cure" people from getting suicidal again. The goal would be much simpler: to help patients develop and internalize new—and person-specific—strategies for coping with future suicidal crises.

Fifth, from the many interviews with patients referred after a suicide attempt I had learned that most of them did not have a useful concept of what had happened to them during the acute suicidal crisis. We needed to teach them about mental pain and the suicidal mode, about altered brain function under emotional stress, about dissociation, and about the long-term risk. I had learned in the past that, to my surprise, people were very receptive to information on brain function.

Sixth, as learning always involves the growth of new synaptic connections, we needed to make sure that our patients were actively engaged in the therapy. Ladislav, as he often did, had a brilliant way to put the problem into words: "The point is: For treating an illness you want a passive, bearing patient, while for treating suicidality, you need an active participation of the suicidal person." In our 1997 paper, I found that right from the beginning we had been quite enthusiastic about the potential of our therapeutic model, although at that time we were far from any proof: "Hypothesis Nr. 5. Interventions based on an action theoretical model can be expected to be effective in the prevention of further suicidal acts."[1]

Seventh, as in a review paper[2] we had found that "outreach elements," such as sending regular letters or even postcards, can have a remarkable long-term effect on suicidal behavior, it was clear that regular letters after the individual face-to-face sessions had to be part of the therapy program. The concept of sending letters goes back to Jerome Motto, a dedicated psychiatrist from San Francisco. I had met Motto at the 1990 Montreal IASP congress, where he presented his work. Motto sent letters to patients who had been admitted after a suicide attempt but had refused any follow-up care. He mailed letters once a month for the first four months, every two months for the next eight months, and every three months for the next four years. The letters read: "It has been some time since you were here at the hospital, and we hope things are going well for you. If you wish to drop us a note we would be glad to hear from you." He compared the letter group with a no-letter control group. The results after two years were remarkable: The suicide rate of the control group was nearly twice as high as that of the letter group.[3] The effect was somewhat smaller after five years. Later, the "caring letter" follow-up intervention was repeated in several countries (Australia, New Zealand, Iran, England, France), with mixed results. In our new therapy, letters would be a way of extending the therapeutic relationship following the face-to-face sessions, according to Bowlby's secure base concept, with the letters representing a continued lifeline or "anchored" connection to the therapist and the experience of a brief but meaningful therapeutic relationship.

We first presented our ideas of a novel therapy program to a larger audience at the 2009 Aeschi conference as the "Bern Short Intervention Programme." Later, we tried to find a catching acronym. ASSI (Attempted Suicide Short Intervention) soon became ASSIP (-Program).

Our new therapeutic program included the following components:

1. An extensive narrative interview: establishing a collaborative therapeutic relationship.
2. A video-playback session: putting the patient into the role of the outside observer, watching the video-recorded narrative session, and reflecting about the suicidal development.
3. Homework task: a handout serving as collaborative psychoeducation.
4. A jointly developed case conceptualization: putting things into context, identification of specific vulnerabilities and trigger situations.

5. Personalized safety planning: collaboratively developed warning signs and safety strategies, printed and handed out to the patient.
6. Optional: a "mini-exposure session," using video from session 1 for a rehearsal of safety strategies.
7. Regular letters: the outreach element, as a continuation of the therapeutic relationship and reminder of the safety measures.

In psychotherapy there is a golden rule: If the patient's (implicit or explicit) goals and those of the therapist do not match, psychotherapy will not be effective. It is as simple as that. Before the first session, patients would receive a short summary of the ASSIP sessions and, most importantly, a description of the therapy goal ("We cannot cure you from suicidality—but we shall work out together which strategies in the future can keep you safe"). We soon found that patients readily went along with this; for many it was a relief to hear that ASSIP was brief and had a clear structure. I remember a patient saying: "You know, I have gone through many therapies for my suicide attempts, and first thought 'not again'—but when I read the information I received on the ward, I said to myself—well, this sounds interesting, these guys obviously know what they are doing. Let's do it."

THE THERAPY COMPONENTS OF ASSIP EXPLAINED

1. Narrative Interviewing

ASSIP is the only therapy that allocates an entire session for the narrative interview. Over the years, as we accumulated hundreds of video-recorded narrative interviews with suicide attempters, we had ample opportunities to develop and refine our skills in narrative interviewing. A key issue is that therapists believe in the *narrative competence* of their patients. Most patients can give us a comprehensive self-narrative with no or very little support from the therapist. We have recordings of amazing twenty- to forty-minute narratives with in-depth stories of the suicidal development; of thoughts, emotions, and actions; trigger events; vulnerabilities related to the biographical background; early negative and often traumatic experiences; etc.—without active therapist interventions. For most people, giving a comprehensive story is the most natural thing—they are just not used to be invited to do this in a medical setting. They are not used to sitting with a health professional who takes a genuine interest in the person in the patient and who really takes time to listen.

2. Collaborative Video Playback

The video playback (or video self-confrontation) changes the patient's perspective from the inner experience of the narrator to that of the external observer. It is a powerful way of establishing self-awareness and preparing the ground for new strategies to cope with future suicidal crises. When discussing the therapy process factors of ASSIP, I often hear from therapists that they believe that the video-playback procedure is probably the main component responsible for the method's remarkable effect. No other treatment for people at risk of suicide has included this therapeutic technique.

When I read Daniel Kahneman's bestseller *Thinking Fast and Slow* (2011), the book about human errors in decision making, it struck me that the so-called dual processing theory could be relevant for suicide (box 10.1).

Box 10.1: About System 1 and System 2

The *dual processing theory* is a psychological theory. It posits that in decision making we rely on two mental processing systems, which Stanovich and West called System 1 and System 2. The model is as follows: System 1 operates automatically, intuitively, and quickly, with little or no conscious input. This system is reliable in familiar situations, and its decisions are generally swift and appropriate. Driving a car is an example. However, in out-of-the-ordinary situations the system is subject to serious errors. For situations, such as driving a car on snow, we need to engage System 2, which operates with conscious control and concentration. The task of System 2 is to override the automatic decisions provided by System 1 when necessary. This control function is crucial to prevent fatal decisions.

However, in certain situations, System 2, instead of intervening, may endorse risky or blatantly wrong decisions: "It is natural for System 1 to generate overconfident judgements, because confidence, as we have seen, is determined by the coherence of the best story you can tell from the evidence at hand. Be warned: Your intuitions will deliver predictions that are too extreme, and you will be inclined to put far too much faith in them."[4]

As we have seen earlier, it is striking that people who survive jumping from high bridges usually say that right after they have jumped and while falling, they immediately switched back to consciously realizing what they had just done, like Kevin Hines: "This is what I have to do. I am a bad person, I am a burden to my family and friends. . . . I hit free fall, and at that second I knew what I did was a terrible mistake. How can I go back? How can I change it? How can I stop it?"[5]

This is a first-person account of a suicidal person who describes the switch from System 1 (intuitive, automatic) to System 2 (conscious). Why does System 1 endorse a life-threatening action? Here is my answer: Suicide actions have a background, a story that often reaches back to childhood. In the developing years we are exposed to the fact that suicide and suicidal behavior happen in our society. We all know that suicide is an option in human life, and we know what the common suicide methods are. In a study we asked people about their early memories of exposure to suicidal behavior. We made two groups: patients who had attempted suicide and patients without a history of suicidal behavior. The introduction to the interview was: "Imagine a child does not know that people can and do kill themselves. Please try to go back in your life and tell me what experiences you remember where you were confronted with the idea that suicide is an option in human life."

One study participant from the attempted suicide group said: "I was confronted with suicide very early in my life, in the first year at school. The mother of a schoolmate threw herself under a train. Up to then, at the age of seven, suicide had been a taboo theme at home, and this event, I have to say, shocked me quite a bit. Later, I heard that this woman was about to become blind and therefore had escaped from this life, which is somehow understandable."

Others remembered reading Goethe's *The Sorrows of Young Werther* at school. The results of this small study showed that, indeed, people who had attempted suicide recalled significantly more memories of direct exposure to suicidal behavior than the nonattempting group.[6] The interpretation is that early exposure to suicide may be stored in a person's brain as a viable action to end unbearable emotional stress and pain. In this case, System 1 may endorse suicide as a solution because it makes sense for the person, and System 2 does not intervene.

If we apply the theory of the two processing systems to our therapy concept, it is evident that in the self-narrative, patients are basically in the automatic System 1 mode (which is most obvious when we watch the video-recorded interviews). In the following video-playback sessions, patients are now in a very different role. As outside observers they engage the conscious System 2 mode, looking at the System 1. With growing insight and with the active help of the therapist, System 2 can now learn to mobilize attention and intervene when a particular pattern (personal warning signs) is detected. "The way to block errors that originate in System 1 is simple, in principle: recognize the signs that you are in a cognitive minefield, slow down, and ask for reinforcement from System 2."[7] I don't think Daniel Kahneman would object to my application of his theory.

A patient said: "It occurred to me that it is good to hear everything again, the whole story, the whole video. The conscious understanding gets stronger. One can see what one did. I suppressed everything inside me and tried to cope with it in that way. With watching it becomes clearer for me, it becomes clear what I did." In the video playback sessions patients are usually glued to the screen, listening to their own suicide story. These are intense sessions, where patients are invited with some prompting from the therapist to jointly reflect about their story, about how to make sense of what they had gone through. Different to the first session, in the video playback session therapists are far more active, working collaboratively with the patient.

Here is what a patient wrote to me after the second ASSIP session:

> Dear doctor
>
> Since I have seen you I have been feeling unburdened. Although about a week ago I experienced again something like beginning thoughts about suicide, I do feel better than three weeks ago, after the suicide attempt. Since then I also talked more with friends, and I tried again and again to explain what happened.
>
> I feel that the interview, and above all, watching together the video afterwards, gave me very much in terms of working through. Today it is clear to me what a "silly" idea such a suicide attempt, or suicide itself, is.
>
> Again, many thanks!
>
> With best regards
> R.W.

The video playback is a way of accessing the suicidal crisis from a different perspective, with the active help of the therapist as coauthor. Neuropsychological research suggests that memory consolidation is most effective when a content is repeatedly accessed in intervals and from different perspectives.[8]

3. Collaborative Psychoeducation

At the end of the second (or first) session patients receive a four-page handout in which we address some major points we believe are relevant to put the patients' own experience into a theoretical as well as practical frame. I had noticed that when patients returned the handout in the following session, many had scribbled their notes on the edge of the page, commenting on the descriptions of mental pain, autopilot mode, etc. Repeatedly we heard that they had passed the handout on to their family members because they wanted them to understand. For many, the reference to the altered brain function was a relief, reducing the sense of guilt about what they had done. As mentioned earlier in this chapter, I have found that people are surprisingly open to brain research related to suicidality. We later upgraded the handout by including empty spaces for patients to comment. In this way the handout became a homework task, with the aim of inducing an active reflecting about the handout text, leaving room for patients' own experiences and thus turning the handout into a collaborative therapy component. I deliberately chose a rather provocative title for the homework handout: "Suicide Is Not a Rational Act," which refers to the acute suicidal crisis (see appendix 1).

4. Joint Case Conceptualization

The case conceptualization, too, had to be a collaborative project, just as the idea of ASSIP as a therapy with a strong therapeutic alliance altogether was. The therapist makes a draft of the case conceptualization according to the ASSIP model, drawing from what emerged from the two previous sessions, if possible using the patients' own words, taken from the narrative and the video playback (box 10.2). The text is written in the first person "because it is your story."

Patient and therapist are sitting next to each other facing the PC screen. The therapist then reads the draft of the case conceptualization, checking after one or two sentences to see if something should be added or revised. The case conceptualization includes the following points:

- The person's basic needs and their biographical context
- The related vulnerability
- The triggering experience and the emotional response
- The actual and past suicidal thoughts and actions

> ### Box 10.2: Example: Case Conceptualization
>
> Dear Lisa,
> As discussed, here is my attempt to summarize the main points from our conversations on the background of your suicidal crisis. The following text is written in the first person, as it is your story.
>
> I was adopted from an orphanage to a religious family. My parents were always preoccupied with my brother, and I never received any attention as a child. My childhood was characterized by neglect and physical punishment. I was also bullied at school. My parents didn't support me. I often felt like no one loved me and that I don't fit in. I've had a death wish for a long time, and I have always carried it with me. I hoped that things would get better when I left my parents and got married.
> The marriage with my husband was difficult. I often suffered from anxiety and depression. After six years he left me, and I was alone with my son. I was devastated, and this was the first time when I made plans to drown myself in the river. I had nobody to talk

to, and I thought that I couldn't bear this life anymore. When something negative happens, I automatically blame myself. I felt I wasn't good enough, and I felt like a second-class citizen. I was unable to go to work, and my doctor prescribed an antidepressant, which I took for two months.

I then had another relationship. But, after three years, one day when I came home, all his belongings had gone. He didn't even bother to leave a message. In the following days I tried to contact him, but he didn't answer the phone. I didn't go to work anymore, stayed at home, didn't bother to eat or shower. From my workplace, nobody made any attempt to contact me. I was devastated. Life had come to a stop. I then decided to take a taxi and went to his apartment. He was not at home, but, as I still had a key, I went in, and I found that he had gotten rid of everything related to me, gifts, clothes, etc. . . . I felt treated like a piece of trash.

I got into such a strange state of mind, acting like a zombie. I knew where he kept the painkillers, I found two boxes, with some thirty tablets each, and I then took the whole lot, left the place and went to the river. I wanted to die. My thoughts were: I am wrong, useless, things will never get better. I texted a short message to my mother. The police found me and I was admitted to the psychiatric hospital. I was there for three weeks, and they gave me a follow-up appointment to see a psychotherapist.

Lisa—we have seen that it is important for you to feel respected and safe in a relationship. Experiences of being

> abandoned and rejected deeply hurt you and make you feel worthless, blaming yourself. These are dangerous situations, where suicide appears as a solution to end this unbearable mental state. In order to prevent you from getting into the same—life-threatening—zombie-like states of mind again, it is important that we establish strategies to keep you safe in the future.

5. Personalized Safety Planning

The case conceptualization leads directly to the safety planning. At this point in the therapy patients are usually highly motivated to develop safety strategies to be used in case of future emotional crises. First, we make a list of *Helpful Long-Term Measures*. These may include treatment of depression, psychotherapy for trauma, seeking help for alcohol problems or for housing problems, etc. Often, patients want to include personal goals, such as learning to set boundaries, to become more assertive, to accept oneself, etc. Next are the *Warning Signs*. These are crucial for responding early to dangerous situations. Patient and therapist make a list of (1) emotions, (2) thoughts, (3) physiological symptoms, and (4) behavioral responses that are indicators of impending doom. Examples are "I feel empty, numb"; "I blame myself"; "my thoughts go in circles"; "I feel dizzy, restless"; "I withdraw from people." This is where System 2 must become active, slow down, and turn the attention to the safety strategies. We distinguish between *early and late (acute) safety strategies*. Most suicidal patients have in the past already engaged in some helpful strategies, maybe without realizing. Together with the therapist more personal strategies are developed, for instance, taking the dog for a walk, cuddling with the cat, playing the clarinet, going out into nature, listening to a certain song, calling a friend or a family member, calling the ASSIP therapist. *Acute strategies* usually involve emergency professional help, such as calling the crisis line or the hospital or going directly to the hospital emergency room.

The graphic model shows the concept of the ASSIP treatment model (figure 10.1). It starts with the notion that people have biographically determined *existential needs or life goals* (e.g., the goal to have a harmonious and stable family). When an adverse experience (e.g., a threatened separation) meets these personal existential needs, *mental pain may act as an energizer with high physiological arousal.* If mental pain is experienced as unbearable and irreversible, *this may lead to suicidal thoughts and plans as a solution and activate a stress-induced change of mental functioning, the suicidal mode,* which is often associated with *dissociative states in which people act as if programmed by an autopilot program.* The model shows that there is a development from mental pain to action planning to the suicidal mode and to a suicide action. This

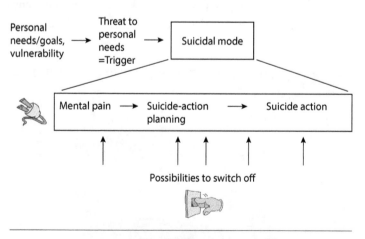

FIGURE 10.1. The ASSIP treatment model.

process will be much faster in persons with a history of earlier suicide attempts, with the suicidal mode activated ("switched on") immediately, as the solution to and relief from the unbearable inner experience. The development may be a matter of minutes, hours, or days. In ASSIP, people learn to identify the personal warning signs for this emotional rollercoaster and to know what strategies help them at different stages of the suicidal development to escape from the dangerous progression into the suicidal mode and the life-threatening action. It is crucial that people learn to use these strategies before it is too late.

Once the safety strategies are completed, the therapist prints the jointly developed case conceptualization. Patient and therapist sign the document as a therapeutic contract. The warning signs and safety strategies are copied into a credit-card-sized concertina-style folded leaflet (which we call leporello) and, together with the signed case conceptualization, handed to the patient (figure 10.2). With the patient's consent, other professional helpers involved receive a copy of the case conceptualization and safety planning.

In a TV documentary about ASSIP I interviewed a patient I had seen three years ago. When I mentioned the folded leaflet,

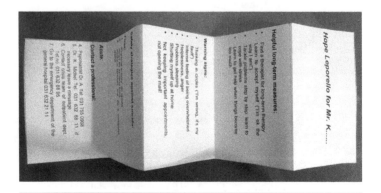

FIGURE 10.2. The credit-card-sized safety plan.

he opened his wallet and proudly showed he still had it ("of course I always keep it on me!"). More recently, we have started to use a smartphone application instead of the folded leaflet.

6. Optional Mini-Exposure

The aim of this session is the rehearsal of the safety strategies developed. This therapy component has been derived from the work of Gregory Brown and others who used role-play for the rehearsal and reinforcement of the safety strategies.[9] In ASSIP we use an action-related narrative sequence from the recorded first session, asking the patient to stop the video at the moment when he or she would now stop the development toward suicide by using one of the agreed safety strategies. Clinical experience has shown that some patients may benefit from this technique. In addition, it is a test of the usefulness of the safety strategies. It is not rare that after the exposure, the safety strategies need to be revised.

7. Letters

Sending regular letters over a two-year period, signed by the ASSIP therapist, is a way to continue a therapeutic, albeit distant, relationship, offering a therapeutic anchorage consistent with the secure base concept. The letters are signed by the ASSIP therapist, and patients are reminded of their direct access to the professional care system. Patients receive semistandardized letters every three months in the first year and twice in the second year. Each letter has a slightly different message, reminding people of the jointly developed safety strategies. Here is the first letter (box 10.3):

Box 10.3

Dear []

Three months have passed since your last appointment with me. I hope that things are going well for you. Let me remind you that if necessary you can contact us at any time. In case I am not available, please ask to be put through to our emergency team or emergency doctor [a telephone number is included here]; they are available 24 hours.

Should things get difficult again for you, please remember the safety strategies we have developed. You must not let things get out of control so that suddenly you only see the one solution. Life is too valuable. As we believe that it can be lifesaving for patients to know they can contact someone, you will receive a short note from us from time to time over a period of two years.

I would be pleased to hear from you if you would like to write a few lines, but you are not required to do so.

You will receive a next letter in 3 months' time. If you no longer wish to receive our letters please let me know by phone or by email.

With best wishes,
Konrad Michel, MD
[signature]
konrad.michel@upd.unibe.ch

About one-third of the patients send us feedback by email, which will then be briefly acknowledged in the next letter. We take this as a proof of the long-term aspect of the therapeutic relationship (after three face-to-face sessions!). When I met an

ASSIP patient in town, she said to me: "Doctor—I am waiting for your letter. Do you know that it is overdue?" I mumbled something about holidays and workload . . .

Here is an email I received from a patient I had seen for the three ASSIP sessions:

> Dear Dr. Michel
> Thank you for your letter.
>
> Over the last weeks I have given up all hope. I stopped seeing my therapist and decided to finally kill myself. Two weeks ago I opened my mailbox because I wanted to delete all messages from my account that I don't want to be read by other people after my death.
>
> Then I found your letter you had sent by email. It left me confused. I was surprised because I had thought that my decision was clear and definite, and now I suddenly felt torn between life and death.
>
> Still, there is no day without the thought that I have to kill myself. I know all the high bridges and buildings, and in the forest I keep a rope in a place I often visit. I know it is my fault that I got into this dead-end from which there is no other escape than death.
>
> Can you understand this? I would appreciate your thoughts.
>
> Kind regards, P.

Of course, I answered the letter and asked her to come to the clinic. She started again with regular outpatient treatment. I couldn't cure her suicidality, but she eventually started to invest in life-oriented activities, and with time the suicidal thoughts lost their urge. She retrieved her rope from the forest and burned it.

The follow-up contacts reach patients with an increased suicide risk: Suicide risk is particularly high after psychiatric hospital discharge.[10]

Is a three-session therapy a therapy? When we presented our new treatment model in international meetings, there were always some skeptical colleagues: "You cannot call a three-session therapy program a therapy—if anything, you may call it an intervention! Therapies require time to build a therapeutic relationship." Although for the sake of the acronym we used the term "intervention," *ASSIP is a therapy*. We have a therapist-patient engagement with a true therapy process.

Let me once more refer to Kandel (see chapter 4): "When a therapist speaks to a patient and the patient listens, the therapist is not only making eye contact and voice contact, but the action of neuronal machinery in the therapist's brain is having an indirect and, one hopes, long-lasting effect on the neuronal machinery in the patient's brain; and quite likely, vice versa."[11] The neuroscientist speaking: even a brief, focused, intensive, and truly collaborative therapy does change the "neuronal machinery of the patient's brain."

We only know if a new idea is any good when we put it to the test, that is, perform a study in which we compare a treatment group with a control group without this treatment (box 10.4). When my colleague Anja Gysin-Maillart joined me, she very soon decided that ASSIP would be the topic for her doctoral thesis, and she took the initiative to develop a study protocol. We were lucky that our department supported the project so that we were able to see our patients—we needed sixty for the ASSIP therapy and sixty for the control group—as part of our clinical work. And, to prove the long-term effect, we decided on a follow-up time span of twenty-four months. It was Anja's idea to contact a publisher to see if they would be interested in bringing out an ASSIP therapy manual. I thought there'd be no chance a publisher would be interested in a treatment manual without evidence of its effectiveness. I was wrong. Huber/Hogrefe in Bern were interested right away. This meant that we had the therapy manual right from the beginning of our study, with detailed and practical descriptions of each session.[12] The English version followed later,[13] and then, finally, in May 2016 we published the randomized controlled trial in *PLoS Medicine*, a high-impact medical journal.[14]

Box 10.4: The ASSIP Randomized Controlled Study

One hundred and twenty patients referred after a suicide attempt were included in this study. With a follow-up period of twenty-four months, the study was running for altogether four years. The patients were randomly assigned either to the control group (60) or the ASSIP group (60). We hypothesized that ASSIP would significantly reduce the number of repeated suicide attempts during follow-up. As mentioned throughout this book, a history of attempted suicide is the main risk factor for suicide and suicide reattempts. All patients had treatment as usual, which could mean anything from psychiatric inpatient treatment, to short-term crisis treatment, to outpatient care, to no treatment at all. Detailed information on treatment as usual at referral to the study and during follow-up was collected, including medication, days of hospital care, number of outpatient sessions, etc. ASSIP was provided as a brief "add-on therapy." The control group, in addition to treatment as usual, had just one single suicide assessment session. All patients were asked to fill in a set of questionnaires every six months during follow-up.

At the twenty-four-month follow-up, the participants of the control group had reported a total of forty-one suicide reattempts, compared to five in the ASSIP group. When we looked at how many participants had "relapsed," that is, had attempted suicide again once or more, the statistician came up with 26.7 percent and 8.3 percent, respectively. For the comparison of the outcome of the two treatment conditions over time, there is a sophisticated statistical procedure called survival analysis (here meaning how many "survived" without reattempts), a figure that we had found in comparable studies. We came up with an 80 percent risk reduction in the ASSIP group. Brown et al., in their 2005 study with ten or more therapy sessions, had reported a 50 percent risk reduction over eighteen months; Rudd and colleagues, in their 2015 study of army personnel receiving twelve or more treatment sessions, achieved a 60 percent risk reduction over twenty-four months. This meant that our ultrabrief therapy was not only as good as the other established treatments but had done remarkably better. Simon Schwab, our brilliant statistician, analyzed the data several times, but the results were robust. We looked at further outcome measures. Two of them were especially interesting: Over the twenty-four months, ASSIP patients spent 72 percent fewer days in hospital (which included medical emergency admissions as well as mental health care). And higher scores of patient-rated therapeutic alliance in the HAq were related to better outcome in suicidal ideation. That is, the more helpful the patients found the therapist in the three sessions, the fewer suicidal thoughts they reported during the first year's follow-up.

The paper went through a thorough reviewing process, which took several months, until it was published on March 1, 2016, in *PLOS Medicine*. The same day I was contacted by newspaper journalists who wanted to report on the results. Some articles had rather special titles: "How Old-Fashioned, Pen-to-Paper Letters Could Help Pull People Back from the Brink of Suicide" (*Washington Post*, April 7, 2016); others were straightforward: "This Simple Therapy Technique May Help Reduce Suicide Risk" (*HuffPost*, April 13, 2016).

In summer 2016 I met with my old friend Terry Maltsberger in Boston, the great Aeschi supporter, a few weeks before his death. Over dinner he suddenly said, "I don't believe it." I asked what he meant. "I don't believe that with three sessions you can have such an effect." I didn't want to argue with him. I believe that the amazing results of our ultrabrief therapy were, still, after years of collaboration, incompatible with his strong psychoanalytic background. And, scientifically speaking, I knew that our ASSIP study results needed to be confirmed in replication studies. I also know that, generally, the results of the original authors of a new therapy program cannot be achieved by other researchers.

Several months after the publicized study, I received an email from Simon, our statistician, telling me that someone from

Stanford University had contacted him, asking for the entire data set so that they could reanalyze the data. He had to enlighten me about the Stanford group, which was led by John Joannidis. To be picked by Joannidis is the nightmare scenario for RCT (randomized controlled studies) authors because he often determines that the conclusions drawn from these studies are unjustified or just utterly wrong. Nevertheless, I agreed with Simon that we would transmit our complete data set. The reanalysis of the ASSIP RCT was then published in the *British Medical Journal* together with other studies.[15] The Joannidis team used a slightly different method for the survival analysis and came up with 79 percent risk reduction (compared to our 80 percent), with the same statistical significance! I told Simon that he should take his wife out for the most expensive meal in the most expensive restaurant in town, on me.

After a presentation of the ASSIP study at an international congress in Oviedo, Spain, a Korean researcher from the London School of Economics contacted us because she wanted to do a retrospective cost-effectiveness analysis of the ASSIP study. She came to Bern twice, and I spent a full day at the LSE with her, working on the manuscript. Our communication at times reminded me of the Tower of Babel syndrome—a psychiatrist and a health economist trying to find a common language. The study was finally published in 2018, with—as expected—very good results for ASSIP.

> This study found ASSIP to be cost saving in reducing the number of suicide attempts in the study population. The results suggest that ASSIP can be considered for scale-up in various local and national settings.
>
> Most of the cost savings were driven by significantly lower costs for general hospital services and less use of inpatient care services in the ASSIP group.[16]

These studies indicate an amazing effect of our ultrabrief therapy on reducing reattempts, and at less cost, with results not matched by any other formalized therapy. However, I need to point out again that the results of any clinical trial need to be confirmed in replication studies. The iron rule is that a treatment can only be called evidence based when other studies in different settings have found it to be effective. This will be the interesting question. Is the treatment transferable to other health care settings, other patients, other cultures? The main publication generated a lot of clinical and research interest. Soon, ASSIP started to spread without much active input from our side. In 2020 there were ASSIP teams in Finland, Lithuania, Sweden, the United States, Austria, Germany, Switzerland, Belgium, and Australia. The ASSIP manual for clinicians has so far been translated into Finnish, Dutch, Korean, and Farsi (Iran). An ASSIP version adapted for adolescents (AdoASSIP) is currently in the pilot phase in several clinics in Switzerland.

Trained ASSIP therapists, once they come to grips with narrative interviewing and with the novel case conceptualization, are enthusiastic about the new therapy. Steven Moore, a psychologist from Syracuse, New York, wrote: "I went to the ASSIP training somewhat skeptical that a three-session intervention could make much of a difference. After completing over twenty cases I have become a true believer in ASSIP and I hope to incorporate ASSIP into my practice in the future."

To be honest, at times, still today I am amazed by the fact that in the ASSIP group, only five people out of sixty "relapsed" (reattempting suicide within twenty-four months). It means that the vast majority had managed to use the warning signs and safety plans we had jointly developed. In other words: The three sessions had an amazing long-term effect. How can this be explained?

In psychotherapy, the elements that are believed to be crucial for a successful therapy are called therapy process factors. ASSIP is a package of several factors we have come across in the previous chapters, such as the novel suicide-as-action model, the actively engaged patients, the video playback, and the clear structure and goals of the therapy. One aspect I consider to be extremely important is that ASSIP respects personal autonomy. Throughout the sessions ASSIP uses a strictly collaborative approach in which the therapist holds back with advice and instructions. People don't change their minds because we tell them to do so. The general principle is that people need freedom of choice. Thaler and Sunstein define the "nudge" as "to alert, remind, or mildly warn another." We can nudge people to adopt a different behavior—if we can convince them that a new behavior would indeed be better for them.[17]

SUMMARY

Treating suicidal individuals needs a specific therapy. We have seen that ASSIP, a brief and novel treatment program that evolved from over twenty years of development and collaboration with others, was found to be very effective in reducing the risk of suicide reattempts. ASSIP does not replace any necessary long-term treatments, psychotherapy, or pharmacotherapy.

Yet even if the model is clinically useful, the big question for me remains how we can reach the vast number of people who are at risk but do not normally seek help. We need to find ways to reach them—by disseminating a new model of suicide. A model that makes sense to people.

11

NOW WHAT DOES THIS ALL MEAN FOR SUICIDE PREVENTION?

While I was writing this manuscript, the parents of a forty-eight-year-old man, married with two children, who had died by suicide came to see me. He had thrown himself under an intercity train a month ago. The son was a successful lawyer in a government position and dedicated father. In his childhood he had experienced two serious suicide attempts by his father, an HR manager in a big firm. The father had an extremely high sense of responsibility and reliability—which at the age of fifty-eight had got him into an existential crisis where he made a serious suicide attempt. The son, like his father, was reliable, responsible, and highly respected as a lawyer, which got him into a top position. When his boss, who had personally supported him in his career, retired, the work situation changed. The new head of the department took his work ethics for granted and burdened him with more and more tasks. He began to work harder in order to gain her appreciation. He developed sleeping problems, problems with concentration, came home late from work, lost energy, and withdrew from his family. His wife and his parents worried about him, yet he insisted that everything was OK. One morning, instead of going to work, he went to the railway track and ended his life.

Such tragedies should not happen.

I so much would have wished that this man had known the signs of a developing disaster, the dangers of developing burnout and depression, the growing tunnel vision that seemed to leave no room for alternative possibilities to defuse the situation, so that important life goals—family, wife, and children—fell out of view. And I wished he could have understood that when his brain started to produce concrete plans when and how to end his life, he still could have gotten off the emotional rollercoaster heading toward the deadly crash. From the outside, it looks so easy: He might have turned to the family doctor, to a crisis center, or a crisis line. Or a close friend. Or his wife. "Listen, I must tell you something . . ." Or that his wife could have found a way to approach her husband (see appendix 3).

Some 50 percent of people who die by suicide do not seek help before their death. Each year thousands of people die by suicide because they do not think that they need medical treatment or that a doctor's office or a mental health service would be the right place to get help. There are many misconceptions about suicide.[1] With our new knowledge about suicidal development, using the frame of a suicide-as-action model, I see a true potential for many people to learn to become more mindful of suicidal thoughts and plans, understanding that there is a danger when mental pain and emotional stress increase, so much so that we may lose control over our actions. In other words: To reduce suicide in this world, it is crucial that the public learns to understand suicide and suicidal behavior as a life-threatening behavior with a personal psychological background. A life-threatening behavior that can be averted.

Too many patients drop out of follow-up treatment after an attempted suicide within a short time. The risk factor model does not help the suicidal individual understand the personal significance of suicidal thoughts, and it does not help prevent future suicidal self-harm. A suicide-as-illness framework seriously impedes therapists in meeting the needs of the suicidal person. When the professional helpers' understanding of the problem and that of the patients diverge, no meaningful therapeutic work will develop—a Tower of Babel syndrome. We came across this problem in several examples mentioned in this book: the violin maker who dropped out of treatment after I told him he was suffering from depression and needed to take an antidepressant, the Afghanistan veteran who flushed the pills he was prescribed own the toilet.

Clinicians and researchers have spent a lot of brainwork, time, and money to identify suicide risk factors, from neurotransmitters in the brain to emotional neglect in childhood. In 2018 over three thousand articles were published, the majority focused on epidemiology and the medical model of suicide.[2] I am far away from discarding the medical model. I am a psychiatrist who takes psychiatric assessment seriously and who prescribes drugs where indicated. I put a lot of emphasis on collaborative prescribing, explaining to the patients why medication is important, taking account of the patients' attitudes toward medication, including earlier experiences (theirs and others'!) with psychotropic drugs. I explain what we know about the mechanism of the effect and possible side effects, and I follow patients closely in adjusting the dosage.

But this book is about the suicidal person in the patient. I have argued that the medical model does not help the suicidal individual understand their own suicidality, to know when things are getting dangerous. The medical model does not look at pain and shame, the main drivers for suicidal behaviors, and it does not explain how changes in brain function can become life threatening.

What, then, must be done? We need to meet suicidal patients with a person-centered approach. We need to relate to "the person in the patient." Suicide and suicide attempts are goal-directed actions with a personal background and a personal meaning. Even in a severe depression, we must try to understand what went on in the person's mind so that they turned against themself. I claim that in each case there is a personal logic that therapists and patients can collaboratively unveil.

From my work with patients I have learned that thoughts about suicide as an option in human life are not uncommon and I would even say not pathological.

Acting on suicidal thoughts is a different matter.

In the narrative interviews I learned that people who had attempted suicide didn't really understand the full significance of their suicidal action, which could have ended in death. Until now, most models of suicide have badly neglected the aspect of dissociation and the concept of the suicidal mode, as described earlier in this book. Very few people—including health professionals—know that when emotional stress and mental pain dramatically increase and when the emotional brain takes over, we may lose the capacity to rationally decide and act. We may later regret what we did—if we get that chance (remember Kevin

Hines!). Furthermore, people do not know that after a first suicidal crisis they will now live with an increased risk of suicide, probably for the rest of their lives. Nor do they usually have a plan for how to cope with it should it happen again. When I first used the psychoeducative handout for patients "Suicide Is Not a Rational Act" (see appendix 1), I was surprised what an effect it had. Obviously, most patients had read it attentively, and quite a number had scribbled personal comments on the edge of the paper, often with exclamation marks. They told me that they had shown the handout to their families because they thought it explained so well what they had experienced, and they wanted others to know what happens to people in an acute suicidal crisis. And, in addition, it helped them reduce the guilt and shame they felt.

Here is what I wish to tell these hundreds of thousands of people who may be at risk, people who would not usually seek help, those who have survived a suicidal crisis, those who have lost a close person to suicide, as well as the rest of the population who so far have not even considered suicide for themselves. I know, it sounds grandiose. I am neither a megalomaniac nor a missionary. I am simply arguing that the broader public needs a new understanding of suicide. Here is the content of my message:

1. In emotionally difficult situations, when faced with an impasse for which we see no solution, it is not uncommon to develop thoughts of suicide as an option. Fortunately, it usually is a long way from suicidal thought to suicidal action. But, still, I want people to know *that when emotional stress increases, our*

brain may trick us into believing that suicide is indeed a viable or even the only solution.

2. *Mental or psychological pain can grow into an extremely unpleasant and disturbing sensation.* It can create catastrophic thinking—"it will never subside, it will even get worse."

3. *The situation may escalate and become life threatening as the emotional stress increases and the mental pain becomes unbearable.* Suicide risk is high when we begin to develop concrete plans about how to end our own life and when we start believing that others (family, friends) would be better off without us. Withdrawing into our narrowed-in world and pretending that everything is OK is another warning sign.

4. *In such moments, our brain is subject to losing the capacity to think rationally. In the alarm mode, our rational thinking and problem-solving capacities are largely abolished.* Some researchers have used the term "cognitive myopia," meaning an extreme short-sightedness in our thinking, not allowing any other considerations. The emotional brain takes over, driving us toward an immediate solution to end the unbearable and existentially threatening mental state. In brief: In highly stressful situations, we cannot trust our brain always to act in our interest.

5. From the accounts of people who survived, *we must assume that the vast majority of people who die by suicide would in retrospect regret what they did.* This is illustrated by people who jump from bridges and survive, saying that seconds after they had jumped, they realized that what they had just done was a fatally wrong action.

6. Finally, people *must* know that suicide is far from being inevitable. On the contrary, *we can learn to become aware of these dangerous developments and to know what to do about it.* The figures of the repeated suicide attempts in the follow-up of the

ASSIP study speak a clear language: Five suicide attempts in the ASSIP group in the twenty-four months after versus forty-one attempts in the control group (chapter 10).

Most of us have in the past experienced minor or major emotional crises that we managed to overcome. Most of us have developed strategies to regulate our emotions. The threat to the self may seem unbearable, the situation hopeless—but we have learned that usually it will subside. There are hundreds of strategies people use for self-regulation, to reduce emotional stress and inner tension, sometimes with help from others. But, once we are on the suicide track, we must know the warning signs, and we need to know what to do to be safe, *before* the suicidal mode strikes and the rational brain falters.

Some people might object, saying that the decision to end their lives was well thought out and rational. I can very well see that my model, which focuses on the acute suicidal crisis, may not do justice to people suffering from serious and possibly hopeless conditions, say, terminal cancer. I assume the same is true for suicide during times of war, imprisonment, political or religious prosecution, etc. The Indian farmer who lost a full year's crop and sees himself in a trap of rising debts would probably leave me helpless as a therapist. The same is true when I read about the shockingly high suicide rates of young people among the Inuit and other indigenous populations.[3] Obviously, my model of suicide cannot do justice to the enormous number of people who end their lives as a consequence of systematic traumatization. Unnecessary to say—it is incredible what members of the species *Homo sapiens* can do to one another.

Yet even if I consider the limitations of my model, let me summarize the message I want to disseminate:

WHAT YOU MUST KNOW WHEN YOU THINK OF SUICIDE AS A SOLUTION

- A deeply painful experience can trigger a state of alarm with high emotional stress.
- Emotional stress seriously impairs our ability to think rationally.
- When we see no solution to the painful experience, there is a risk that the emotional brain takes over, driving us toward an immediate solution to end a seemingly unbearable mental state.
- From people who have survived a suicide attempt we know that in retrospect they realized that they acted as if driven by an autopilot program, that they were not their usual selves.
- Being ashamed and withdrawing into ourselves can seriously damage our life.
- The main and first remedy is: Talk, talk, talk. Seek help.
- You may also watch this three-minute educational animation by Science Animated SciAni UK, at https://konradmichel.com/#video.

A CASE FOR IMPROVING SUICIDE HEALTH LITERACY

The U.S. Department of Health and Human Services (HHS) defines health literacy as "the degree to which individuals have the capacity to obtain, process, and understand basic health information needed to make appropriate health decisions." Health literacy empowers the individual by helping them understand the basic health information necessary to make appropriate health decisions. Health literacy means that we know what to do when we have a high fever and a dry cough (COVID-19 symptoms!), sudden chest pain, dizzy spells, unusual headaches, etc. To promote health literacy, the message must be easy to read and understand. Today, much health-related information is easily accessible through the internet. Most internet users report that they use search engines for get health information; such searching is the third-most-popular web activity, after checking email.

The U.S. National Action Plan to Improve Health Literacy is based on the following principles:

- All people have the right to health information that helps them make informed decisions
- Provides everyone with access to accurate and actionable health information
- Delivers person-centered health information and services
- Supports lifelong learning and skills to promote good health[4]

I assume that even in remote places today people know that severe pain, persistent high fever, or a snake bite leading to inflammation are signs that they should seek professional help. What if people learned that repeated or persistent thoughts about suicide may herald a crisis that could be life threatening? What if they knew that suicide is not "normal" and that most people who attempt it regret it? That when the stress level increases they may find themselves in a situation in which they lose control over their thoughts and in which the brain may get them to do things they would not normally do?

A web search for health literacy and suicide yields interesting results. Some links lead to depression and suicide, which we will not pursue here. Other, mostly recent articles, demonstrate that suicide literacy is a highly complex issue. What health professionals believe is good for the broader public may not be good or even be harmful: "Targeting the broader public to raise awareness of the scope of the problem may adversely affect vulnerable individuals. Adverse effects may be due to an increase in norms that describe suicidal behavior as common or frequent. This may increase the likelihood that individuals will believe that engaging in suicidal behavior is widespread and therefore acceptable."[5]

Indeed, the question is how to talk about suicide. Suicide literacy must aim to change attitudes and beliefs and increase knowledge about suicide. But what kind of knowledge? Scott J. Fitzpatrick argues that

> suicide literacy research presupposes that health professionals are a source of unassailable knowledge when it comes to suicide and its prevention. A growing body of international literature on service users' experiences of clinical services following a suicidal act has shown that medical interventions are often viewed as discriminatory, culturally inappropriate and incongruent with the needs and values of people who are suicidal. These incongruities between the knowledge, attitudes and language of experts and those with experiential knowledge of suicide represent more than individual deficiencies. They are illustrative of the power, values and interests that challenge current suicide prevention practice. At issue here is the relationship between medicine and experiences of distress and suicide.[6]

In addition, improvements in health professionals' knowledge, attitudes, communication, and clinical skills are needed. This, however, is an issue by itself and will be addressed in chapter 13.

Results of suicide literacy projects are mixed. Chan and others assessed students' help-seeking intentions. They found that those who normalized suicide had significantly lower intentions of seeking help for thoughts of suicide.[7] Maria Ftanou and others investigated the effect of public service announcements on eighteen-to-twenty-four-year-old students, using three interventions: "Talk to someone" (talking to others can help those who may be thinking about suicide), "find what works for you" (to find strategies that could help one cope with suicidal thinking), and "life can get better" (emphasizing that over time, stressful situations can improve and negative feelings can reduce). One might have expected a difference between the intervention and the control group. There was none.[8]

An interesting project is the Literacy of Suicide Scale (LOSS), aimed at gaining information on the public's knowledge of suicide and its association with help seeking.[9] The scale has been included in several studies across nine countries with general community, student, and clinical populations. The scale can be used to quantify levels of suicide literacy, as well as to test the effectiveness of interventions designed to improve suicide knowledge in the community. The same authors found that more severe suicidal ideation was associated with higher glorification of suicide and that males had lower levels of suicide literacy.[10]

Obviously, to improve suicide literacy, the challenge is to find the right messages, use the right channels, and reach the right populations. For instance, men and boys should learn to express emotions in ways that are perceived as strengths rather than weaknesses. Suicide literacy research should above all focus on groups known to show more resistance toward help seeking. When we think of addressing people who are at risk, one suicide-specific problem that is hardly mentioned in the literature may be that suicidal individuals do not necessarily think that they have a problem and that they need help. As we have seen earlier in this book, suicide as a solution makes sense in the suicidal persons' special personal suicidal logic. They tend to feel that there is nothing wrong with thinking of suicide—given the difficult situation, the emotional stress they are in, the abysmal shame, and the loss of identity. Here, the "don't do it" approach will not work.

Knowledge about suicide that is meaningful to individuals at risk is crucial. Yet we must respect people's autonomy, even when we believe we know that suicide is a wrong decision. Let me again refer to Thaler and Sunstein's notion of *libertarian paternalism:*

> The libertarian aspect of our strategies lies in the straightforward insistence that, in general, people should be free to do what they like.
>
> The paternalistic aspect lies in the claim that it is legitimate for choice architects to try to influence people's behavior in order to make their lives longer, healthier, and better.... In many cases, individuals make pretty bad decisions—decisions they would not have made if they had paid full attention and possessed information, unlimited cognitive abilities, and complete self-control.[11]

The question is whether the suicide-as-action model can be helpful to increase help seeking. It differs from other suicide models with its focus on the acute presuicidal development, with mental pain and high emotional stress, where normal life-oriented goals can suddenly go out of focus. It integrates knowledge from brain research and explains why suicide in this context *is* taking action but taking the *wrong* action. The main messages are that people need to be aware of the danger and that they can learn to change the course of events and regain control. The model is easy to understand, and it increases empowerment. Importantly, it is *not a theory to explain* suicide but a *model to understand* the suicidal mind. It goes beyond the relationship of suicide and suicidal ideation with the stigma-associated illness model and its association with psychiatric pathology and psychiatric treatment.

Print and electronic media are powerful means to reach a large population and introduce new knowledge. For example, Jason Cherkis, an American journalist, got hooked on Jerome Motto's prevention project of sending letters to people who had attended an emergency department after a suicide attempt. Cherkis wrote a very engaged and touching story about Amanda, a twenty-nine-year-old woman, her therapist, and the importance of keeping in touch through caring letters, a means of reducing suicide initiated by Jerome Motto in San Francisco. He got interested in our brief therapy and sat in with my colleague Anja in Bern when she was doing her ASSIP therapy sessions. The impressive article was published in the *Huffington Post* on November 15, 2018, accompanied by a YouTube video.[12] The article generated many positive reader comments: "This is a wonderful article. I hope many people read it"; "This is the most wonderful thing I have ever experienced in my life"; "This was an incredible piece, I've been feeling suicidal most of this year due to my anxiety." After two years, the article had more than 750,000 views. It included a link to our book chapter "The Narrative Interview with the Suicidal Patient,"[13] which, within a short time, was downloaded over three thousand times. Who would have expected this? Unfortunately, we don't know who the readers were. I assume that covering suicide carefully, probably when connected to personal stories, has the potential to change public misconceptions and correct myths, which then will encourage those who are vulnerable or at risk to seek help.

A project I admire is LivingWorks ASIST, a remarkable low-threshold nonmedical suicide first-aid training.[14] It is a training program that in a two-day workshop enables a wide range of potential helpers to intervene effectively to help people experiencing thoughts of suicide. The emphasis is on teaching suicide first-aid to help a person at risk stay safe and seek further help as needed. Participants learn to use a suicide intervention model to identify people with thoughts of suicide, seek a shared understanding of reasons for dying and living, and develop a safety plan based upon a review of risk. Over one million people have been trained in LivingWorks suicide intervention skills, and the basic assumptions are very much in line with the underlying model of suicidal behavior in this book: suicide as such is not mental illness; thoughts of suicide are understandable, complex, and personal; and most people with thoughts of suicide want to live. LivingWorks also offers safeTALK, which in a four-hour workshop teaches participants to recognize when someone is thinking about suicide and connect them to an intervention provider, enabling "everyone to make a difference," from neighbors, to taxi drivers, and to barbers (men!), etc.[15]

SUMMARY

The medical model does not help people understand the personal dynamics of a life-threatening suicidal crisis. The suicide-as-action model offers a person-centered model of suicide and has been shown to be extremely effective in reducing the risk of reattempts after an ultrabrief course of therapy. The model does not explain suicide as a human phenomenon but has been developed as a model that helps suicidal people understand their vulnerability toward certain adverse experiences, the role of trigger events, and the concept of the suicidal mode's sudden changes in mental functioning, in which the emotional brain may drive us into a life-threatening action. We need to clearly distinguish between suicide risk factors such as depression and the suicidal individual's personal dynamic.

I believe that the suicide-as-action model has the potential for changing the public's views on suicide and suicidal behavior. Experts agree that we must improve suicide health literacy. Public campaigns must be carefully targeted and closely followed and monitored. One aim should be to improve help seeking. This, however, requires that health professionals and counselors meet the needs of suicidal patients. In this respect, I believe, we have a long way to go, although there are some encouraging projects.

12

A SPECIAL CONCERN: YOUNG PEOPLE

I know how it is to lose a child.

Our son, Alex, age twenty, ended his young life by suicide on November, 25, 2001.

Alex was a tall, good-looking, bright young man with a promising life ahead of him. After high school he had done his fourteen weeks of army training, then began his studies at a Swiss school of economics. He lived in a student apartment together with friends. He was uncertain about his goals; in fact, he told me that he didn't see any goals in his life. He had gone to high school "because it was the natural thing to do," being the son of parents with an academic background. He said he didn't know if the school of economics was the right thing for him. He had chosen a prestigious university because some of his friends had applied there. He didn't see himself earning money to own a family home (like his father), nor did he want to put his efforts into a professional career (like his father). I told him that there were no expectations from his parents to continue his studies and that he could do whatever he wanted. I realized how difficult it was for me to really understand his problem. For me, to go to high school and medical school had been a privilege. I was the first person in our family history to go to university. Alex failed the exam after the first year. We, his parents, assured him that he could take all the time in the world to find out what made him happy. To complete the picture, I need to add that one year ago, Alex had attempted suicide while both parents were away. He was admitted to psychiatric inpatient care for a few days. He claimed the reason had been a breakup with a girlfriend. On my recommendation, he began psychotherapy with an experienced psychiatrist in the town where he lived and studied. As his father, I couldn't be his therapist.

One Saturday when he was expected to come home for the weekend, he was missing. He didn't answer the phone. The police found him three days later in the forest. He had hanged himself.

Three weeks later I was to appear as the keynote speaker at a national congress. I introduced myself as a father in a state of shock. I felt that it would be wrong, as an expert on suicide prevention in my country, to go into hiding. In the first weeks after Alex's death I felt that I couldn't continue with life at all. I used to say: "If one of my beloved children should die, I couldn't imagine how life could continue." Yet I survived. We were lucky to get the support of professional colleagues and a particularly engaged psychologist and dear friend. Colleagues at work told me that I should stop seeing suicidal patients. It took some time before I learned to separate my personal life from my professional identity. Personally, it felt like walking on thin ice over deep and dangerous dark water that could devour me. Over the years the ice got thicker, and I began to feel safer again. I am a father who lost his son by suicide and a psychiatrist involved in clinical work and suicide prevention. I decided that my work had become an important part of my identity and that I needed to continue with it, even seeing suicidal patients or parents who are mourning the loss of a child. After a presentation in a congress a group of bereft parents approached me. They first hesitated, but then they told me that it was a relief for them to hear that the catastrophe could also happen in a family of an expert in suicide prevention.

Parents very often noticed that their child was not well but found no way to get their son or daughter to open up. In my very first study, in which I interviewed families of one hundred individuals who had made a suicide attempt or had died by suicide, I heard many stories, but the ones that were most difficult to bear were the stories of families who had lost a child through suicide. In contrast to the adult group, none of the young suicides had shown the clear-cut picture of a depressive syndrome. Very few of them had given warnings of the impending disaster. From these interviews with parents I concluded: "(1) adolescents and young adults who commit suicide rarely show signs of a psychiatric disorder prior to their death; (2) these young people rarely talk to someone about their acute emotional conflicts and suicidal plans; and (3) most of these suicidal acts show a great degree of determination and are characterized by the use of means that leave little chance to survive."[1]

Alex had said he would come home for the weekend. For Saturday morning he had made an appointment with the local barber. But he never turned up that day. Like many other parents we will never know what went on in his mind on Friday night. He left no letter. Nothing. For months, we parents tormented our brains to understand, but to no avail. We developed all kinds of theories and turned in circles until we had to accept that we shall never know.

Since then, I have seen many bereft parents. When trying to understand what might have driven their son or daughter into death, the key issues that emerged appeared to be somewhat similar. Remember the story of the eighteen-year-old Oliver in chapter 7. He got himself into a deadly trap by not telling the parents that in his vocational training he had stopped attending classes, and, what was worse, he had blatantly lied to his parents about the thesis paper he was due to submit. He had not even started to write a first draft. He died the day the truth would have come out into the open.

Another young man told me:

> For a long time I had built a construct of lies, I had actually lied to my closest people. More and more I realized that the moment would come where my construct would become shaky. And I started to think that putting an end to my life would be easier than unwrapping everything and telling people that I had lied to them. Well, I have always thought that I had this option to end it all, and I knew that there would be my army gun, which would be a safe method.

Annelie Törnblom and her colleagues from Stockholm in 2013 interviewed fathers and mothers of thirty-three boys and young men who had died by suicide.[2] These were impressive in-depth interviews of three to four hours. Almost all parents reported that they had worried about their boy. That the boy wouldn't talk or even tried to avoid direct contact with them. The parents often knew that the boy had been exposed to stressful or even traumatic events or that he had problems at school with friends and grades. Parents typically described their boy as hiding behind a mask. He would not reveal any of his problems. When the parents tried to confront him, he would fool them by flatly denying any problems. He could present misleading plans, tell the doctor he would never consider suicide, or take a morally wise position with friends: "I'd never do this to anybody." In trying to understand the suicide of their sons, the core issue that came out from these interviews was *shame*: "Shame for what the boys had done. Shame for what happened to them. Shame of physical appearance. Shame for who they were."[3]

Wikipedia defines shame as a very "unpleasant self-conscious emotion typically associated with a negative evaluation of the self, withdrawal motivations, and feelings of distress, exposure, mistrust, powerlessness, and worthlessness." In adolescence, the sense of self is a very delicate thing. To develop the self and a sense of identity is an enormous task. Most of us have had periods where we felt that something was not right with us, that we were not like others, where we blamed or even hated ourselves. Maybe you remember how you tried to conceal your deficits and your shame. Adolescents tend to hide their unprotected self behind a mask. This, however, may become life threatening when the "no-problem" mask grows into a false self, a dead end from which it is difficult to back out of.

Insight gained: Shame is a frequent driver for young people to hide their inner turmoil behind the no-problem mask.

Box 12.1: About Shame

Shame is a normal feeling. We don't usually show it, we like to keep it to ourselves. Shame and pain belong together. Young people are vulnerable. Adults appear to be strong—with the emphasis on "appear." For a young person it is absolutely normal to have episodes of self-doubt, feeling inadequate, or even hating oneself. You may think that there is something seriously wrong with you. That you will never be like your parents. It is normal to keep these painful feelings to yourself. Typical reasoning is that it would be dangerous to tell your parents, because if they didn't understand, it would be proof that there is really something wrong with you. Instead of risking not being taken seriously, it seems safer to keep it secret. Yet the longer we hide our secrets from others, the bigger the barriers to finally talking to someone become. It needs a hell of a lot of courage!

Confiding our secret thoughts and feelings about ourselves to someone is the solution. Talking to a person we trust—whatever we are so ashamed of—will not make things worse! Experience shows that grownups will acknowledge your courage to talk about your secrets. It will be a relief for you—and your parents.

For some people, it may be easier to talk to a professional person. For others, it may be easier to chat with an anonymous counselor. You may use one of the links listed further along in this chapter. The important thing is to talk about it, to put the thoughts, which threaten to destroy us, into words. Everybody needs trusted people who will listen to us. Counselors and therapists are used to all kind of stories and are nonjudgmental in their work. And you can rely on their confidentiality.

This is the way to take away the burden. Test it.

Shameful thoughts, some examples:

Having failed an exam but at home have said otherwise

Having dropped out of college

Having stolen

Having lied (or not told the truth)

Having had an affair and worrying about HIV

Having had a negative sexual experience

Having unusual sexual desires

Feeling as LGBTQ+
Hating one's body
Being rejected by others
Whom can you talk to?
A friend
Parents
General practitioner, family doctor
Teacher
Counselor
Priest
Crisis line
Crisis chat
Online—helpline
etc.

The big question for us adults is: How on earth can we get young people to see that shame can never be a reason to end a young and promising life? How can young people learn that making mistakes and feeling ashamed are the most normal things in human life? Cate Curtis from Waikoto University, Hamilton, New Zealand, asked university students about their views of help seeking: "It's seen as a weakness. Suggesting it can be a slap in the face; it depends on the approach"; "It's hard to reach out and if someone suggests it, you think you must be crazy"; "I wouldn't see a counselor or whatever unless I was absolutely desperate. Imagine if someone found out or they told your parents! They would just go on and on about it. Your friends would think you were mental."[74]

> **Box 12.2**
>
> If you are a young person and have an idea about how we can get young people to recognize the problems of shame and of hiding behind a no-problem mask—and what they could do about it—please go to konradmichel.com and let me know.

After a suicide or suicide attempt of their child, parents are left devastated with feelings of guilt, thinking that they had failed to understand and thereby save their child. There are endless questions for which there are no answers. "Where had we missed something? Was it our fault that our child did not confide in us? Should we have insisted when our son or daughter flatly denied any difficulties?"

As a personal message to bereaved parents, let me here include to a letter I wrote to the Editors of *Crisis, the Journal of Crisis Intervention and Suicide Prevention*, three years after the death of our son. The letter was a response to an article that had been published in 2004 ("Denial in Suicide Survivors"), written by a prominent suicidologist and professor of psychology.[5]

> Dear Sir,
> Dear David,
> I guess a "Short Report" in *Crisis* need not be evidence-based. Yours certainly isn't. The message as I understand it is simple: Survivors ignore the cues of impending suicidal action in their significant others, therefore, they are protagonists in that person's suicide. You cite the 1959 Robins study where 69% of the suicides retrospectively were found to have given prior

communication about their suicidal intent—so, you assume that these signs had simply been ignored.

Everyone involved in the prevention of suicide knows how extremely difficult it is to correctly interpret possible signs of risk. Have you ever told schoolteachers what the warning signs of a risk of suicide are? Their answer usually is: "This is so unspecific that at one time or other most of our kids must be considered suicidal." Only 9% of teachers and one-third of high school counselors thought that they could recognize a student at risk (King et al., 1999).[6] Most adolescents do not disclose their suicidal ideation to their parents or teachers but to their peers (Pronovost, 1990).[7] . . .

Recognizing a risk of suicide is not enough. The real difficulty is to know what to do when a suicide risk is suspected. Do you, David, know what to do if a worried parent asks you for help in the case of their adolescent who is taking all the posters and decorations off the walls of his room and cleaning his room (your example), but who flatly refuses to see any therapist? Do you know the answer? Still, let's assume you are successful in getting this young man to come to see you. What then? Nearly 75% of the people who end their lives through suicide have seen a medical professional within their last year (Miller & Druss, 2001). I strongly agree with what Ian Fawcett said in the 2003 AAS meeting: "Some people think all we have to do is to get people into treatment, but the fact is that we are not very good at treating suicidal people."

Our son was very reluctant to see a psychotherapist, but eventually did so. The therapist was an experienced senior colleague—but he wasn't able to stop our son from ending his life a few months after he had started therapy. Other mental health professionals who are survivors have similar stories (e.g., Anonymous, 2001).

David, your "Short Report" is embarrassing. I agree with you that for the communication of suicidal ideation we have to look at the sender as well as the receiver of such messages. However, accusing survivors of having *ignored* (your italics) signs of impending disaster doesn't get us anywhere. For a report in a professional journal the real questions to be asked are: Why do so many people who are at risk not seek professional help? Why do so many people drop out of treatment? Why do so many people kill themselves in spite of being in treatment?[8]

As we shall see later, so far there are practically no treatments for suicidal adolescents that have been shown to be effective in reducing the risk of suicidal behavior.

The young, particularly young men, in the transition to adult life have a high suicide risk in most Western countries. So far, prevention programs have not led to a measurable reduction in suicide rates among adolescents.[9]

In 2021 I started working with the dedicated group around Gregor Berger, MD, at the Zürich University Hospital of Child and Adolescent Psychiatry, on a project to adapt ASSIP to this age group ("AdoASSIP"). The therapists video-record each session for me to supervise them. The young patients are girls and boys, aged thirteen to eighteen, referred after a suicide attempt. It is striking, sometimes shocking, to hear how many of these young patients talk about suicide as if it was the most normal way to cope with difficulties and unpleasant feelings (interestingly, a recent Chinese study found that, similarly, a weak sense of life's value was common among students, and over 60 percent thought that suicide was a way to end or evade problems).[10] When we asked our young patients to tell us what occupied their minds, they often mention "being wrong," not fitting in with peers or siblings, and not being acknowledged the way they are by their families. It slowly dawned on me that these young people are in the middle of the difficult process of developing their identity—this is different from adults, who talked about adverse experiences, which meant that their sense of identity, already formed, was at stake. For many young people, a suicide attempt does not appear to be something they regret, nor do they really seem to think that safety plans for the future are that important. It was soon clear that we had to add a fourth session, in which we included parents and, if possible, long-term therapists involved in the aftercare. In the family session, adolescents usually read the AdoASSIP case conceptualization and safety plans to their parents. This is a powerful procedure: the adolescent has the floor, and the patients are the audience. Parents are often visibly struck when they hear their child's emotional problems laid out before them. Often there appears to be a huge gap between the hidden needs of a child and the parents' perception that their child is doing well. Here we see again the unfortunate tendency of young people to keep their problems to themselves, hiding behind the everything-is-OK mask. The family session brings this out into the open and opens up new ways of family interaction. We had expected that our young patients would refuse to read their story they had approved in the third session to their parents, and we first thought to leave the decision to do a family session to the young patients. I am still thankful to Anna-Lena Hansson from Malmö, Sweden, with whom I discussed this question. Her answer was: "Konrad, with your background of attachment theory, you must know that these young people need their families—whatever the situation is at home—and you cannot leave the decision to include their parents or not to your young patients, who have just survived a suicide attempt!" Right she was. So far, I don't think the Zürich team has had anyone who refused.

Working on this project, it became obvious what a huge demand for a specific therapy for young suicidal people there is. The highly motivated team under Gregor Berger can barely cope with their many referrals. However, coming from my experience with adults, I am still puzzled by the attitude of young people toward suicide. One explanation is that adolescents' sense of a personal identity and of the continuity of life are hardly existent. However, here again, wouldn't talking to parents normally be the way to deal with problems?

> **Box 12.3**
>
> If you are a young person, I would like to know how for you suicide became a way to cope with negative feelings about yourself and what helped you deal with it without harming yourself. I would very much appreciate your answer. Please go to konradmichel.com and let me know.

Young people are exposed to all kinds of threats to their fragile selves. Bullying is one of them. It is done with the intention to intimidate and distress the victim. It is a form of rejection by peers, leading to isolation and a breakdown of self-esteem. Bullying is a way to cause feelings of inadequacy, shame, self-depreciation, self-blame, and of "being wrong." Bullies have easy game in targeting the weak spots of their peers. Cyberbullying is a particularly powerful form of bullying because of the bully's ability to remain anonymous, and, as cyberbullying takes place in a digital world, it has the potential to reach a large audience. For the victim, it doesn't need much to get the—quite normal—negative thoughts about oneself to develop into an emotional rollercoaster. Even the most glamorous teenage girl may feel that she is indeed the ugliest person on earth. I remember talking to an adolescent girl admitted to the emergency room after she had taken an overdose, telling me that her boyfriend had suddenly left her for another girl. This made her feel "*dumped like piece of trash, not treated like a human being.*" The shame of being abandoned, the pain of being rejected. My spontaneous reaction to such a case usually was: "Why on earth does she blame herself? Shouldn't she be furious with this guy who dumped her like this?" Of course, as a good therapist I kept this to myself. A therapist doesn't tell people how they should feel.

In social media it is impossible to avoid suicide as a topic. Facebook claims to have developed an artificial intelligence program to identify social media users at risk of suicide via their online posts, yet it is not clear whether such interventions are effective. In July 2019, a Danish influencer with more than 336,000 followers posted a suicide note on Instagram. It was taken down by the family after two days, but it sparked a debate in Denmark on how to monitor online content from influencers. The Danish minister of children and education stated that influencers had an "editorial responsibility" in line with the standards of the "old press." And: "We have a society where the mass media of today has changed and the standard of mass media communication has to change and has to apply to the new mass media. It's different media but the same ethics."[11]

The Australian #chatsafe project (https://www.orygen.org.au/chatsafe) provides excellent guidelines for peer-peer communication in many different languages, helping and encouraging young people to talk safely about suicide on social media. Statements that provide vulnerable people with information about how to end their life, suicide notes, or goodbye notes must clearly be avoided. In the #chatsafe intervention campaign adolescents received messages once a week for twelve weeks. The participants' self-assessment questionnaires showed that the intervention improved aspects of online behavior, with participants being less likely to share suicide-related content, more likely to monitor their posts for harmful content, and more likely to contact someone directly if they believed they were at risk.[12] The Australian Orygen website contains many clinical resources and interventions for at-risk young people (https://www.orygen.org.au/Research/Research-Areas/Suicide-Prevention).

LGBTQ+

Mental health problems are frequent among LGBTQ+ youth. In Switzerland, 1,108 adolescents from ages seventeen to twenty completed extensive questionnaires about sexual attraction, suicidal ideation, and self-injury.[13] Twelve percent of the female participants and 4.4 percent of the male participants reported identifying as LGB. Twenty-eight percent of LGB adolescents reported a history of suicidality (ideation, plans, and attempts) in comparison to 12 percent in the sample of heterosexual youth. We know that sexual-minority individuals are more exposed to stigmatizing experiences and emotional distress. Emotional stress is often related to the suicide of others and the lack of social support in coping with bereavement.[14] A meta-analysis found high rates of suicidal behaviors in adolescents and adults with gender incongruence, that is, a persistent incongruence between an individual's experienced gender and their assigned sex.[15]

On the one hand, it seems that in recent years there has been a development toward depathologizing sexual minorities. On the other hand, we have to be careful about the influence of media reports on the relationship between LGBTQ+ identity and suicide risk. Silvia Canetto argues that *"the assiduous media reporting of LGB youth suicidality, and the dominance of struggle, suffering, and vulnerability themes in LGB youth"* has led to a certain normalization of suicidality in LGBT youth.[16] It would not be helpful if the LGBTQ+ community viewed suicidality as a "normal" response to life problems. Here, we have to be aware of the influence on attitudes toward suicide in vulnerable populations that electronic, print, and social media can have. I have written earlier in this chapter about my experience in the Zürich "Ado-ASSIP" project, where many of these young patients talk about suicide as a seemingly normal way of coping with difficulties of not belonging.

THE COVID-19 PANDEMIC

The COVID-19 pandemic interfered heavily with the needs of teenagers in developing a sense of identity. For the lucky ones, their support networks have been enough to get them through the pandemic, while for others, the abrupt changes in daily life, with months of virtual classes and social isolation, took a toll. What used to be familiar and give structure to their lives crumbled. Young people need peers to maintain a sense of self-worth and to develop their sense of identity. Being cut off from friends and activities left them deprived of necessary social interactions. It led to increased levels of stress, panic attacks, mood swings, sleeping problems, personal insecurity, and suicidality. Virtual school classes didn't replace that necessary feedback. Grades dropped, and many started to withdraw, feeling increasingly worthless. According to the *CDC Weekly* for June 18, 2021, from February 21 to March 20, 2021, emergency department visits for suspected suicide attempts were 50 percent higher among girls ages twelve to seventeen than during the same period in 2019; among boys in that age group, visits increased 3.7 percent.[17] Symptoms of depression and anxiety disorders doubled during the pandemic.[18] Similar changes have been reported from other countries. The new situation was a challenge to ER staff, who were not prepared for this dramatic increase of young people in crisis. In the University Hospital for Child and Adolescent Psychiatry in Zürich, the number of the staff in the clinic for emergency admissions had to be increased dramatically. A specific training program for therapists with online supervision was introduced. Interestingly, reports show that the rise in emergency admissions for mental health problems had started before the onset of the pandemic.

The crucial issue is how to get young people to seek help. Teenagers and adolescents tend to think they should be able to cope on their own. They fear that seeking help will create more problems for their parents. If anything, they prefer to talk to friends. Young men are particularly reluctant to seek help. The suicide-related communication of boys, compared to girls, is often ambiguous or diluted by "humorous" connotations.[19] In Western societies—and this still appears to be the case in the twenty-first century!—the male identity is largely defined by not admitting anxiety and problems with oneself. Expressing pain and emotional sensitivity is not cool, nor are feelings of weakness, uncertainty, helplessness, or sadness. They may show irritability and hostile-aggressive-abusive behavior. In an inquiry, young men indicated that the driving force was losing hope of performing according to their own goals, combined with an unwillingness to disgrace themselves by revealing their shortcomings.[20]

When adolescents visiting a pediatric clinic were asked if they thought that doctors and nurses should ask kids routinely about suicidal thoughts, surprisingly, 88 percent answered yes. Only few thought it was "kinda weird." It appears that adolescents may be more willing to disclose suicidal thoughts to a health care provider than to their parents.[21] But many adolescents are very sensitive about the clinician's attitude:

> Being a good listener is one of the vital things . . . and the way they sit as well. The way they sit or look at you, their gesture is very important. You don't want to be ignored at that moment when you're telling your story.
>
> When they're typing, you kind of feel, "What are they typing? Now I feel like maybe I shouldn't say that. What if they type it into the system?" . . . things like that.[22]

However, in an emerging suicidal crisis many of them will not contact a professional helper but will instead search for relevant information about emotional and mental problems on the internet, where they may also be attracted by websites promoting and encouraging suicide. That said, there are lots of excellent internet sites for young people offering help, including chat services where users can stay anonymous and where they are free to decide how much they want to disclose.

Box 12.4: Helpful Online Resources for Young People (and Others)

The U.S.-based Zero Reasons Why (https://www.zeroreasons why.org/) project includes a community initiative that involves a teen counsel and lots of personal stories. There are also excellent parent resources and useful links.

An attractive Australian site is the Suicide Call Back Service, https://www.suicidecallbackservice.org.au/, providing professional 24/7 telephone and online chat or video counseling. You can download a Suicide Safety Plan app, and there is practical advice for what to do if you are with a person at risk. Another excellent site that offers resources for suicide safety planning is Beyond Blue (https://www.beyondblue.org.au/get-support/beyondnow-suicide-safety-planning).

Another example is the Finnish chat platform www.sekasin.fi/, staffed by professional helpers. In 2018 the platform had some 1,400 inquiries daily, with 98 percent of users being under thirty years of age. I particularly liked the welcoming sentence: "Welcome to Sekasin's Chat! Everyone is sometimes confused. Come to chat." And it continues: "In the chat, you can discuss anything that burdens your mind. We will listen and support and advise you if necessary. Our goal is for you to be heard and met by a person. All kinds of emotions and subjects are allowed, and we don't feel that sorrows may be too small or too big."

I was impressed when I read about the Crisis Text Line Project (www.crisistextline.org). Crisis Text Line offers free, 24/7, high-quality text-based mental health support, provided by volunteer crisis counselors trained in using active listening, collaborative problem solving, and safety planning to help texters in their moment of crisis. It is wonderful to see that so many volunteers are ready to go through a thirty-hour-long training and then spend four hours a week (mostly at night) reflecting listening to people seeking empathy and help. The topics range from abuse to bullying, anxiety, and gender and sexuality to eating disorders, depression, suicide, and self-harm. The suicide section provides excellent advice for people contemplating suicide and for those concerned about someone else. Even if you are not suicidal, test it; it's impressive!

The Virtual Hope Box from the National Center for Telehealth & Technology (https://apps.apple.com/ch/app/virtual-hope-box/id825099621) is a well-designed smartphone application containing a large number of tools to help users cope with stressful situations. Users can add helpful contacts. The tool that appealed most to me is the "Controlled Breathing" exercise.

There are school-based awareness programs aimed to improve knowledge, attitudes, and help-seeking behavior. An interesting project is Saving and Empowering Young Lives in Europe. As part of this project, the Youth Aware of Mental Health Program (YAM) is an intervention that addresses young people's negative perceptions of themselves and aims to improve the skills to cope with adverse life events and stressors, which often trigger suicidal behavior. In five one-hour classroom sessions over three weeks, young people are active participants. With the support of the instructor, everyone works to understand different perspectives and come up with possible solutions to problems. In a study, YAM was effective in reducing new cases of suicide attempts and severe suicidal ideation by approximately 50 percent. New cases of depression were reduced by approximately 30 percent in the youth participating in YAM. I highly recommend the free-access paper by Wasserman and colleagues. It addresses the problems that may occur in the interaction between adult professionals and adolescents. Teens aged fifteen to seventeen were asked for their opinions regarding the program, and the answers give a picture of young peoples' attitudes vis-à-vis mental health issues.[23]

Applications targeting safety planning are mostly used in combination with treatments for suicidality. The apps allow users to save their personal lists of suicide warning signs, create coping strategies, identify positive contacts and social settings to distract from the crisis, identify family members and friends available to help, find professional help and resources, and make their environment safe from lethal means that may be used in a suicide attempt. The use of safety apps is not as easy as one might expect. When young users of the Danish MYPLAN safety planning app were asked for feedback, some felt that the app does not make it easier to reach out and ask for help "because of pride and/or not wanting to worry others" and that the app would be hard to use when users are in the "red zone, and that it can then be hard to motivate oneself to continue to use the application." The involvement of professionals in the safety planning is therefore considered to be important.[24]

I like the idea of involving young people in projects for the young. The Hong Kong video project is a well-done co-creation suicide prevention video produced with a popular YouTuber.[25] Will adolescents watch a twenty-minute video? I am not sure. An interesting approach is to let the students produce their own videos. In such a project, the young people said that they had learned a lot about warning signs and how to help suicidal people. The students also said that they were surprised to learn how much incorrect knowledge on suicide and suicide prevention was present.

Is there anything parents can do? I often hear that if young people do talk to someone, they usually prefer peers. However, a survey found that young people do actually mention parents as key people in the help-seeking process.[26] Parents, on the other hand, often feel uncertain how to respond, they are uncertain about how best to support a young person who is engaging in self-harm, and they may not manage to initiate professional help for their child. An overview of studies exploring young peoples' views on what parents could do to help came up with a number of key messages. The one I like most is: "Family Context: Young people wish for more love, attention, time, support and care from families. Parents should take an active interest in the young person and their life, as well as trying to make the young person happy and understanding the challenges that young people face."[27]

A problem arising between parents and their suicidal adolescents is that parents tend to attribute interpersonal reasons to the suicidal behavior, while the young see suicidal behavior more as their own inner experience.[28] Surprisingly—but not really!—parents very often are not aware of their children's prior self-harm behavior. When parents attribute their child's self-harm to problems in growing up and in developing a sense of identity, this may lead to more supportive parenting. However, when self-harm is perceived as "naughty," "bad," or acting out against the parents, this will obviously not help the child feel understood.

Neither will it be helpful simply to attribute the child's suicidal behavior to mental illness.

An interesting approach is the Youth-Nominated Support Team Intervention for Suicidal Adolescents–Version II (YST-II), which focuses on building and strengthening a youth-nominated and informed network of supportive adults. Adolescents nominate trusted adults from their family, school, and community whom they believe would be helpful in a supportive role following the adolescent's hospitalization for suicide risk.[29]

What is deeply worrying are the increasing suicide rates in girls aged ten to fourteen,[30] which stands in contrast to the usual gender distribution. I found a possible (and convincing) answer in the most informative book *iGen*, by Jean M. Twenge.[31] Twenge uses the term "iGen" for those born between 1995 and 2012. It is the first generation that grew up with cell phones, internet, and social media. She looked at large, nationally representative surveys of some 11 million Americans. By about 2012, she noted large, abrupt shifts in teens' behavior and emotional states. Depressive symptoms among tenth graders increased in parallel with smartphone ownership—and far more prominently among girls. One study found that the hashtag "#selfharmmm" zoomed from 1.7 million mentions on Instagram in 2014 to 2.4 million in 2015. The suicide rates in the United States have risen dramatically since 2007. It is more pronounced for girls, with three times as many suicides in the twelve-to-fourteen-year-old girls in 2015 than in 2007. Social media use is more strongly associated with cyberbullying and with emotional problems in girls than in boys.[32] Victims of cyberbullying have a higher risk of suicidal behavior.[33] There are many such tragic cases. One of them is the moving story of Amanda

Todd.³⁴ In Korea there is the story of Sulli, twenty-five, a successful Korean pop star. Sulli became the target of online abuse for not wearing a bra, for livestreaming an innocent night out drinking with a friend, for addressing older male colleagues by their first names, and—her most egregious "crime"—for describing herself as a feminist.

Problematic online behaviors include sexting, cyberbullying, and internet gaming—all associated with a higher risk of self-harm among adolescents. There is a relationship between the time spent using social media and mental health problems. Time spent on social media increases the risk of being a victim of cyberbullying. Facebook and Instagram expose adolescents to idealized self-representations that negatively influence body image, particularly in girls. Twenge refers to statements from teenagers in Nancy Jo Sales's *New York Times* bestseller: "They make you think you have to change yourself, like lose weight or gain weight, instead of just being yourself"; "every day it's like you have to wake up and put on a mask and try to be somebody else instead of being yourself, and you can't ever be happy."³⁵ Teenagers get hooked on Facebook and Instagram. They go to bed with these social media, and—if they don't look at their phone during the night—they are back on it first thing in the morning. Thanks to Frances Haugen, we know how Facebook and Instagram get young people systematically hooked on social media use.³⁶ Paradoxically, this constant connectedness makes adolescents feel lonely. Parents who have not grown up with social media can hardly support their children in dealing with the inherent dangers. The U.S. National Crime Prevention Council website contains valuable information for parents on how to address cyberbullying and internet safety with their children (https://www.ncpc.org/resources/cyberbullying/).

Dangers for teenagers also come from movies depicting suicidal behavior. The well-known Netflix series *13 Reasons Why*, released in March 2017, is based on the bestselling 2007 young adult novel of the same name in which a teenaged girl, Hannah Baker, dies by suicide. Hannah is exposed to bullying, rumor-mongering, friends who turn on her, and rape. She leaves behind thirteen cassette tapes that identify the people and circumstances that contributed to her death.

Abuse like this does happen to teens, no question. The problem with *13 Reasons Why* is that the way the story develops is as if "suicide is almost a natural consequence of the traumatic life experiences that she was dealing with."[37] Furthermore, the sheer amount of time it took Hannah to record the thirteen tapes that chronicle the people and circumstances leading up to her death would probably have helped defused her crisis. "If we want to have an honest dialogue about suicide, it should start with something that resembles reality rather than an extreme outlier."[38]

Bridge and colleagues analyzed the increase of suicide following the Netflix series: "This national study identified an increase in suicide rates for children and adolescents aged 10 to 17 years after the release of the first season of the Netflix series *13 Reasons Why*. We estimate that the series' release was associated with approximately 195 additional suicide deaths in 2017 for 10-to 17-year-olds."[39]

The first investigation also looked into activity on Twitter and Instagram, two of the most popular platforms among U.S. teens. They retrieved 1.4 million tweets and 26,322 Instagram posts that mentioned the series between April 1 and June 30 in 2017. Social media attention peaked in April and faded after June—a window of activity that coincided with increased suicides. An article published in 2020 sums it up: "*13 Reasons Why*: The Evidence Is in and Cannot Be Ignored."[40]

Personally, it goes beyond my understanding how the makers of *13 Reasons Why* ignored the Suicide Reporting Recommendations (https://reportingonsuicide.org/), which are based on nearly forty years of scientific investigation into the effect of portrayals of suicide. An early and strong example is the German TV series of a nineteen-year-old male student who throws himself under a train, which was broadcast twice, in 1981 and 1982. There was a 175 percent increase of suicides in that age group, by the same suicide method, over seventy days after the airing of the episode.[41] I don't think it is asking too much from filmmakers planning a movie that includes a person's suicide to be aware of and take into account the evidence-based contagion effect. The decision to remove a graphic scene from *13 Reasons Why* comes some two hundred unnecessary teenage deaths too late.

Portraying suicide as a natural consequence of negative and traumatic life experiences sets an engram (a memory trace) and a model with a lasting effect for a whole generation. As discussed in chapter 2, setting a model for young people who can identify with the protagonist is called the "Werther Effect," which takes its name from the 1774 book by Johann Wolfgang von Goethe, *The Sorrows of Young Werther*. Werther, a young artist, dies by suicide following an episode of unrequited love. The book was very popular and led to some of the first known examples of copycat suicides, with young men dressing the same way as Werther and ending their lives with the same method. The theory of social learning[42] describes the causal link between media coverage of suicide as a behavioral model and suicidal behavior in a society. The effect is strongest when the model and the recipient share the main personal characteristics (age, gender, rejection, failure, unhappiness, etc.). Exposure of the audience can lead to adopting the television's consensual view of society.

When suicide methods are prominently depicted, this should be seen as "natural advertisements for suicide."[43]

SUMMARY

In many countries, the numbers of young people dying by suicide have been increasing in recent years. We have seen that youths (often male) have an unfortunate tendency to withdraw and to hide behind the "no-problem" mask. I have seen many caring and warm families where the parents had not noticed anything special in their child or, in those who had, where they had found no way to look behind the boy's (or girl's) mask. We have seen how important it is to understand that children and teenagers often feel insecure, inadequate, unlovable, or simply a failure. How teenagers have not yet developed a sense of personal identity and continuity of life and how difficult it is for parents to understand the needs of their children.

Yes, I failed with our son. Or, let me put it differently: We failed to communicate. And we got no second chance.

Today, it is disturbing to see how teenagers and adolescents are influenced by the use of social media and Netflix. Portraying suicide as a normal behavior is a form of social learning with a long-term effect. Describing a suicide method is an advertisement for that method. In *Contagious*, a book I highly recommend, I found the following sentence: "If you want people NOT to do something, don't tell them that lots of peers are doing it."[44] This, by the way, is the reason why in this book you'll find very few suicide statistics.

13

FOR HEALTH PROFESSIONALS: IT'S ABOUT THE PERSON IN THE PATIENT

If we are to reduce the burden of suicide, new approaches, new directions, new collaborations are needed.
—Berman and Silverman, "A Call to Clarify Fuzzy Sets"

The need for innovation in our approach to understanding and caring for suicidal patients is pressing.
—Stanley and Mann, "The Need for Innovation in Health Care Systems to Improve Suicide Prevention"

This chapter has been written for health professionals. "The person in the patient"—the phrase referenced in the title—was coined by Dr. Alex Sabo, a dedicated psychiatrist from Pittsfield, Massachusetts, at the 2015 Aeschi conference in Vail, Colorado. I believe that most therapists would agree with me that in clinical practice, and particularly in psychiatry, the art of *good clinical practice* is to connect with the *person*, the individual, the human being—as opposed to a routine, one-size-fits-all approach to the suicidal *patient*. And I believe that a suicide-as-action framework can provide the conceptual basis for a person-centered understanding of suicide and a meaningful therapeutic relationship.

Most physicians have an imprinted startle response when they suspect that a patient may have thoughts or plans to end their life by suicide. For the average medical professional, in clinical practice, suicide as a topic is largely terra incognita. In training we learned that acute suicide risk requires admission to a psychiatric institution, often against the patient's will. Before my metamorphosis to the suicide-as-action model, I perceived suicidal patients as driven by some inaccessible mysterious force related to some disorder of the brain.

Physicians faced with suicidal patients have to assess the suicide risk, which is basically an impossible task. As in a somatic emergency situation, it is considered to be the clinician's responsibility to decide on the indicated procedure. This task is invariably related to the fear of a wrong judgment, the liability factor, and the fear of future litigation should the patient die by suicide. In addition, there is the thorny issue of the usefulness of psychiatric hospitalization to protect patients from their self-destructive tendencies. To be on the safe side, clinicians tend to admit patients. In cases of involuntary hospitalization, this may mean involving the police, who will take the handcuffed patient to the emergency room for evaluation and from there to the psychiatric institution. However, the latter are not at all safe places, with increased suicide rates during in-patient care and after discharge. Many patients admitted because of suicide risk in retrospect view hospitalization as a negative, if not a harmful experience.[1]

We shouldn't be surprised that patients are holding back disclosing their suicidal thoughts, driven by shame, pain, hopelessness, and the loss of self-esteem. There is the patients' fear that the physician will be unable to deal with the disclosure of suicidal thoughts and plans; the fear of being "locked up" in a psychiatric institution, treated against their will, probably with strong medication and the fear of losing their job and being stigmatized for the rest of their life. In short, the fear of losing control. Even when people *are* in treatment, many—too many—do not manage to overcome their inner barriers to open up to their family doctor or psychiatrist. It is not at all rare that people kill themselves the day of their appointment with a health professional.

And we shouldn't be surprised by the fears of health professionals to ask patients in a straightforward way about suicidal thoughts.[2] A few years ago, at a congress of general and family medicine, I ran a workshop titled "Don't Be Afraid of Suicidal Patients." Usually, in a medical congress, the topic of clinical (medical) suicide prevention attracts five or six people. This time the room couldn't hold all the people who wanted to attend. My message in the workshops was: Imagine suicidal patients as human beings who have their inner reasons to consider suicide—and about which we can converse with them, the same way we speak with other people, people like you and me. Try to look at suicide as an action with an inner (subjective) logic, an action patients can explain, just as we all explain actions with stories (see chapter 6). Show genuine interest in the suicidal person and try the new role of being an attentive listener, signaling that the patient has to help *you* understand, with you as a health professional in the "not-knowing" position. An unusual role! Try to relate to the "the person in the patient." You'll soon learn that this approach creates a bond between the suicidal person and you and that this will be a basis on which both protagonists can enter a working relationship to collaboratively assess suicide risk, find out together what the patient needs to be safe for the next hours and days, and jointly develop a treatment plan. Later, I ran the same workshop at the annual meeting of young primary care physicians. The workshop was fully booked and had to be done twice.

In my clinical work with suicidal patients I have learned that for a true working relationship we need to be open to the inner world of each suicidal individual. If psychological pain is an issue, then we want to understand what hurts and what is so painful in the person's life. We want to understand the meaning of the triggers of psychological pain and how the person deals with it. And, when we hear about an act of self-harm, we want to understand how things have built up to an emotional crisis that seemed to leave no other option than ending the unbearable state of mind and with it, one's life.

It is not the depression that kills but the person who acts according to a personal logic. "The therapists must sometimes set aside their theory-based assumptions in order to understand the client's subjective experience."[3]

I couldn't agree more.

Following the first Aeschi meeting in the year 2000, the Aeschi Working Group (see chapter 9) formulated guidelines for clinicians, which are still valid.

Box 13.1: The Aeschi Guidelines for Clinicians

1. The goal for the clinician must be to reach, together with the patient, a shared understanding of the patient's suicidality. This goal stands in contrast to a traditional medical approach where the clinician is in the role of the expert in identifying the causes of a pathological behavior and to make a diagnostic case formulation. It must be made clear, however, that in the working group's understanding a psychiatric diagnosis is an integral part of the assessment interview

and must adequately be taken into consideration in the planning of further management of the patient. The active exploration of the mental state, however, should not be placed first in the interview but follow a narrative approach.

2. The clinician should be aware that most suicidal patients suffer from a state of mental pain or anguish and a total loss of self-respect. Patients therefore are very vulnerable and have a tendency to withdraw. Experience suggests, however, that after a suicide attempt there is a "window" in which patients can be reached. Patients at this moment are open to talk about their emotional and cognitive experiences related to the suicidal crisis, particularly if the clinician is prepared to explore the intrasubjective meaning of the act with the patient.

3. The interviewer's attitude should be nonjudgmental and supportive. For this the clinician must be open to listen to the patient. Only the patient can be the expert of their own individual experiences. Furthermore, the first encounter with a mental health professional determines patient compliance to future therapy. An empathic approach is essential to help patients reestablish life-oriented goals.

4. A suicidal crisis is not just determined by the present; it has a history. Suicide and attempted suicide are inherently related to the patient's biography, and the clinician should aim at understanding them in this context. Therefore, the interview should encourage patients to deliver their self-narratives ("Can you please tell me, in your own words, what is behind the suicide attempt"). Explaining an action and making understood to another person what made the individual do that action puts a suicidal crisis into perspective and can be instrumental in reestablishing the individual's sense of mastery.

5. New models are needed to conceptualize suicidal behavior, models that provide a frame for the patient and clinician to reach a shared understanding of the patient's suicidality. An approach that does not see patients as objects displaying pathology but as individuals that have good reasons to perform an act of self-harm will help strengthen the rapport. The most common motive is to escape from an unbearable state of mind (or the self). A theoretical model that understands suicide actions as goal directed and related to life-career aspects may prove to be particularly useful in clinical practice.

6. The ultimate goal should be to engage the patient in a therapeutic relationship, even in a first assessment interview. At a critical moment in a patient's life, meaningful discourse with another person can be the turning point, in that life-oriented goals are reestablished. This requires the clinician's ability to empathize with the patient's inner experience and to understand the logic of the suicidal urge. An interview in which the patient and the interviewer jointly look at the meaning of the suicidal urge sets the scene for their dealing with related life-career or identity themes. The plan of a therapy is, so to speak, laid out.[4]

This is what my dear colleague David Jobes promotes in the Collaborative Assessment and Management of Suicidality (CAMS), a framework for working with suicidal patients, independent of therapeutic orientation (see chapter 9).[5] David played an important role in the development of the guidelines.

More recently, Jacinta Hawgood and others have issued "gatekeeper training standards," including knowledge, skills, and attitudes—quite a bit more sophisticated than my first training program for primary care physicians, published in 1992 (see chapter 2).[6] I am also very impressed by Jo Robinson from the University of Melbourne, Australia, and her team, for the work they do with general practitioners and their approaches to young people at risk:

> Most importantly, having a collaborative dialogue with a young person, over and above the use of any tools or instruments, was viewed as key to the assessment process: It's more important to be able to have a dialogue with your patient and some direct discussion about what's going on in their head, what are they thinking about.[7]

The Australian group offers an online training for general practitioners and other primary care or health care professionals to help support the identification, management, and assessment of young people who may be at risk of suicide and self-harm (https://orygen.org.au/Training/Resources/Self-harm-and-suicide-prevention/Modules/Suicide-prevention-in-primary-care-Best-practice-w).

I am extremely pleased to see that communication between health professionals and suicidal patients has now again become a research topic. I was particularly delighted by the work of Selma Gaily-Luoma and others from Finland.[8] One of their conclusions is: "The responsiveness of services to individual needs and preferences was described as key, with service users emphasizing that one valued intervention modality (e.g., psychotropic medication) cannot substitute for another (e.g., therapeutic conversation)."

The clinician in a busy emergency room faced with an eighteen-year-old girl who took an overdose of painkillers will probably perceive her as rather passive, someone who does not want to talk about her reasons for her actions of self-harm. After all, from the intake notes it seems clear that this girl took the overdose after her boyfriend left her. Most probably she wanted to threaten or punish him or make him feel guilty. Or gain attention. Whatever her reasons, it obviously was manipulative behavior. After adequate medical treatment, she will be sent home to the care of her parents, with a follow-up appointment for psychological assessment, which she will probably not attend.

Imagine another scenario: The clinician sits down with this teenage girl (alone, no parent or boyfriend in the room) for fifteen or twenty minutes, accepts her as a deeply hurt human being, and asks her: "Let me try to understand, I would like you to tell me what lies behind this overdose; I have time, I left my beeper with the nurse outside." Let's assume that the approach is successful (what I would expect) and that in fifteen to twenty minutes the clinician and the patient come to a shared understanding. And both feel calm. Let's assume that the girl talked about her inner pain, which had been unbearable, which she can connect to earlier experiences, not only a previous breakup but also the time her parents split up, when she was seven—and felt it was her fault. The narrative approach establishes a first therapeutic relationship and includes the patient as an active participant in the assessment of suicide risk and the therapeutic procedure. However, good clinical practice also requires an adequate psychiatric assessment—for which the clinician needs to switch into the role of the medical expert. I usually tell the patients that now I will ask a number of questions in order to assess their mental state in the recent days and weeks, to make it clear that I am now in the role of the medical expert. The questions will cover signs and symptoms of affective disorders, medical and psychiatric history, past suicide attempts and suicidal thoughts, self-harm, substance abuse, and the actual psychosocial situation.

A preliminary case formulation will include a "storied description" of the suicidal crisis, the method and medical seriousness of the self-harm action, the major suicide risk factors (above all, prior suicide ideation and attempts, suicides in the family, etc.), and the mental state assessment, with a preliminary psychiatric diagnosis. This is followed by a collaborative risk assessment and the formulation of adequate clinical care, including first safety plans ("whom do you contact in case things get worse?").

Let me try to put this into a short formula. When faced with suicidal patients, clinicians have two roles: first, to try to understand the story of the suicidal development and, second, to assess the patient's mental state and short- and long-term suicide risk and discuss treatment plans. In the first role, the patient is the expert of their story; in the second role, the health professional is the expert.

"Suicide Risk Assessment Doesn't Work"—*Scientific American* ran this title in its March 28, 2017, issue.[9] Declan Murray and Patrick Devitt, two psychiatrists, reviewed several studies investigating the usefulness of suicide risk scales. For instance, an article in the *British Journal of Psychiatry* had concluded in 2016:

> All of the scales and tools reviewed here had poor predictive value. The use of these scales or an over-reliance on the identification of risk factors in clinical practice, is, in our view, potentially dangerous and may provide false reassurance for clinicians and managers. The idea of risk assessment as risk prediction is a fallacy and should

be recognised as such. We are simply unable to say with any certainty who will and will not go on to have poor outcomes.[10]

Lanny Berman, the prominent wise old man of the AAS (American Association of Suicidology), drives it home crystal clear: "Your local weather forecaster is more likely to be wrong than right in predicting the weather a mere 48 hours out; suicidologists cannot be expected to do any better in predicting suicidal behavior, hence we should ban the word from our suicidology vocabularies."[11]

Personally, I have always found it very odd when researchers talk of "suicide prediction." The problem is that suicidal behavior does not fit into the somatic illness model that assumes a linear development from suicidal thoughts to suicidal action. Human actions are determined by individual psychological and interpersonal dynamics. Seemingly banal events may trigger acute suicidal behavior within hours or minutes.

Since the article in *Scientific American*, numerous studies have come to the same conclusion. In 2019, a review of seventeen studies found that "99 of every 100 individuals predicted to die by suicide will not."[12] The idea that suicidal development is a linear process that, with growing severity, will eventually lead to suicide or attempted suicide is a fallacy. The ideation-to-action framework is based on the concept that with the right risk factor variables (which so far research hasn't found) we will be able to predict at what moment in time an individual will move from suicide ideation to a suicidal action.[13] Surprisingly many researchers believe that machine learning, a sophisticated system of algorithms, will allow more accurate suicide prediction:

> It would be ideal to follow large, representative, and demographically diverse samples (i.e., >10,000 individuals, consistent with

sample sizes used in national and cross-national epidemiologic research) over a number of years while assessing a large number of risk factors yearly, or more frequently (e.g., via email/ smartphone). Data from these large-scale samples could be used to examine prospective predictors of the transition from suicide ideation to suicide attempts or completions (e.g., among those without a history of prior attempts), as well as to inform the development of risk indices and algorithms for predicting suicidal behavior, including moderators of multivariate risk index models.[14]

Here we have the old medical model popping up, where prediction is based on the assessment and evaluation of the parameters of a pathological condition, which allows us to make a prognosis of how the illness will develop. This concept treats patients (persons!) as passive objects without goals and existential needs.

David Rudd's Fluid Vulnerability Theory posits that we need to differentiate between a long-term baseline risk that varies from individual to individual (which, for instance, is higher for individuals with a history of two or more suicide attempts) and a short-term risk, which is highly determined by aggravating factors for limited periods of time (hours, days, weeks) and which will then return to the baseline level when risk factors are dealt with.[15] I am not the only one with a critical view on the "ideation-to-action framework," which assumes a linear development from a starting point (suicidal thoughts) to an end point (suicide or suicide attempt) at some later time and which is far away from the clinical reality of the suicidal individual: "To truly understand why Jim's suicide occurred then, we must also consider why Jim's suicide occurred on Monday instead of the day before, the day after, or any other day. In other words, what changed that Jim reached his figurative 'tipping point' only on Monday?"[16]

Indeed, a person's very individual basic needs and vulnerabilities would be difficult to feed to an algorithm. And, how would artificial intelligence foresee the moment in time when an important person (family member, peer, colleague, etc.) will send a text message that triggers an existential crisis? As long as we do not put chips into people's brains, even the most sophisticated AI will not be able to log into the person's most intimate inner experience, a person's thoughts and emotions, which are related to the person's—unique—biography.

Yet in clinical practice, when faced with suicidal patients, clinicians need some risk estimate in order to make decisions about the indicated clinical care of the patient. Here, the Fluid Vulnerability Theory can help the clinician better deal with the situation. We need to distinguish between long-term and short-term risk factors. The main *long-term risk factors* have been known for long: A history of past suicide attempts is the strongest indicator of a long-term risk. After attempted suicide, the suicide risk is, probably lifelong, increased fortyfold or more, it is higher when the suicide attempt was medically serious,[17] and it increases with further suicide attempts. The main other long-term risk factor is a major psychiatric disorder: affective disorders (depression, bipolar disorder), substance abuse, and personality disorders. Other relevant long-term risk factors are past suicidal thoughts and past or recent plans about a suicide method. Algorithms in electronic medical records can help alert clinicians to a long-term suicide risk. For instance, the "High Risk for Suicide List" in the U.S. Veterans Health Administration ensures that a "flag" is placed on the veteran's medical chart to notify all VHA staff to discuss suicide risk factors and prevention measures with the identified patient, ideally at each clinical contact.[18]

Short-term risk assessment is a different matter. However, there are several pitfalls. First, we need to remember that it is easy for patients simply to deny suicidal ideation. Much depends on how clinicians ask about this topic. A negatively polarized question will reinforce a negative answer. Rose McCabe found in video-recorded exchanges between providers and outpatients that 75 percent of questions were negatively phrased, for example: "Not feeling low?" or "No thoughts of harming yourself?"[19] Questions were typically close-ended and designed to elicit a yes/no answer, thus constraining any patient desire to tell their story. The outcome, then, will likely be a kind of tacit collusion between patient and provider that suicide cannot be discussed in a psychiatric interview. This form of avoidant communication has been so wonderfully described as "dancing without touching," meaning that there is a mutual agreement between the suicidal patient and the clinician that it is absolutely OK to avoid the difficult questions.[20] In the McCabe study, only 25 percent of the questions were positively phrased in such a way that an expectation was set for disclosure.

We also need to be aware of typical situations where suicide risk can change within a short time. In inpatient settings, this is the case in the first few days of admission, when patients are on leave (hours, days), and after discharge. Busch and colleagues in their study of seventy-six suicides that occurred while people were in the hospital or immediately after discharge found that all had been assessed for risk of suicide in the week before, yet 78 percent had denied suicidal ideation.[21] I think it is obvious that none of the suicide risk scales would have reliably picked up any of these patients. Some of them may have made concrete suicide plans but flatly denied it, while for others, suicide may in the meantime have been triggered by some new, unforeseen event. No-suicide contracts are not recommended. David Rudd and colleagues have formulated a "Commitment to Treatment Statement (CTC)" as a practice alternative.[22] It is an agreement, signed by the clinician and the patient, with the goal to "assist in establishing and maintaining a healthy therapeutic relationship and collaborative process by articulating the need for open and honest communication and identifying the roles and responsibilities of both clinician and patient in the treatment process in general," facilitating "active involvement of the patient in the treatment process, including readily accessing emergency services when and if needed."

Another period of increased risk is the transition from protected inpatient status back to "real" life, when people are faced with returning to their unsolved psychosocial problems. "Home" is often not experienced as a place of safety, comfort, healing, and tranquility but as a dangerous place psychologically—a return to the overwhelmingly scary "lion's den" of uncertainty and unknown, potential threat.[23] The transition from inpatient care to outpatient follow-up requires particular attention. Unfortunately, the clinical reality is that "people are usually not consulted about the planning of their discharge or care. Indeed, the limited literature shows that a 'top-down' or 'paternalistic' approach is the norm for discharge planning."[24]

There is only one way to deal with the problems in short-term risk assessment: Include the suicidal person as an active participant in risk assessment. Take time, ask patients to help in exploring the risk in a collaborative manner. It is exactly this approach that Tobias Teismann and colleagues from three German universities have proposed. It is an article I always use when teaching in mental health institutions. The short formula of their recommendations is

Risk assessment should be understood as a collaborative process in which the therapist recognizes that he does not have sufficient expert knowledge of a patient's suicide risk potential. . . . With this in mind, suicidal patients should be included in the decision-making process for the further therapeutic measures. It is not the clinician who decides on the basis of an alleged risk categorization, but patient and therapist collaboratively develop a concept for dealing with the suicidal thoughts and behavior in a collaborative process (cf. Jobes, 2016).[25]

True risk assessment can only be done in close collaboration with the patient.[26] This can be done in a structured way, for instance with the Suicide Status Form (SSF) by David Jobes.[27] In CAMS (Clinical Assessment and Management of Suicidality), every session begins with the patient and therapist filling out the SSF form. The patient rates themself on a Likert scale, from 1 to 5 on suicidal markers for psychological pain, stress, agitation, hopelessness, and self-hate, in addition to making a subjective assessment of suicide risk. Patient and therapist work together to try to understand what increases suicide risk and why.

Peterson and colleagues have suggested that patients can be asked to answer some questions on a five-point scale: "How concerned should your therapist be that you might cause physical harm to yourself in the next 2 months?"[28] Here, too, the approach addresses the collaborative context of suicide risk assessment and demonstrates that only the patient can do a tentative estimate of the risk. Patients are made partners through this process and are treated as experts about their own lives. Others looked at the predictive value of the Self-Assessed Expectations of Suicide Risk Scale's questions in young patients.[29] The self-rated risk scale was useful in predicting suicide attempts above clinician-administered assessment of suicidal ideation severity. Questions used were, for instance: "If you have serious thoughts of killing yourself in the future, how confident are you that you will be able to keep yourself from attempting suicide?" or: "If you have thoughts of killing yourself in the future, how confident are you that you will tell someone?"

In medicine there is a general trend toward Shared Decision Making (SDM). For me this simply means good clinical practice. After a presentation at an international congress, where I spoke about narrative interviewing, a rheumatologist asked me: "Do you think I could also use your approach with my patients?" I was rather surprised by this question, and there is only one answer: patients who are active participants in managing their care have generally better outcomes and better treatment adherence.[30]

Of course, it would be unrealistic to assume that a truly collaborative and joint assessment is always possible. In clinical practice the clinician's gut feelings may be indicators of risk. For instance, feeling uneasy and being aware of missing the sense of collaboration and of a therapeutic relationship may be warning signs that I am faced with a patient who is at risk. In many cases of suicide that occurred in the hospital or shortly after discharge, the clinical staff would tell me that they had felt uneasy with this patient, even though the patient had repeatedly denied suicidal thoughts or plans. The clinicians' emotional responses to suicidal patients are the focus of Igor Galynker's clinical research.[31] His research group found an association between clinicians' emotional responses to high-risk patients, specifically conflicting emotional responses, and patients' subsequent suicidal behavior. Galynker argues that when faced with a complex but at the same time rapid judgment, "mental shortcuts" integrating an affective impression will serve as positive and negative tags, consciously or unconsciously, and may help increase the reliability of risk assessment. Risk assessment can be improved by combining established—but unreliable—assessment tools with the Therapist Response Questionnaire-Suicide Form.[32] The negative items in this ten-item questionnaire include: "I had to force myself to connect," "I felt dismissed or devalued," "The patient gave me chills," "I felt guilty about my feelings toward him/her," "I thought life really might not be worth living for him/her." An interesting instrument is the Suicide Crisis Syndrome (SCS), which focuses on the acute presuicidal state of mind and on psychological processes that occur in the days, hours, and minutes leading up to suicidal behavior.[33] SCS has been developed for health professionals, but it comes close to the suicide-as-action model with its focus on the acute suicidal mental state. I can see that SCS could be helpful to get us closer to a common ground between patients and clinicians.

Negative feelings toward suicidal patients are well known in clinical practice. Empathy and therapeutic alliance are often put to the test, depending on the course of a patient's personality and the level of their suicidality. Negative feelings toward the suicidal patient may develop, or, on the other hand, empathy with the suicidal wish may interfere with therapy. John Maltsberger and Dan Buie, in a classic paper, have written about the personal reactions of therapists from a psychoanalytical perspective.[34] Mental health professionals are expected to have insight into their emotional reactions vis-à-vis their patients. These are determined by our own biography. Among my colleagues, once I got to know them personally, I heard many moving stories, including suicides in the family or their own suicidal past.

A book for health professionals I highly recommend is *Becoming a Doctors' Doctor: A Memoir*. Michael Myers is a physician who started residency in internal medicine and then moved on to train in psychiatry. If you have read the previous chapters of my book, it's easy to see why I recommend this book: "Your patients are much more than their medical symptoms. You have to understand the context of their lives and how much factors like fractious or distant relationships with family members, unemployment, loneliness, religion, ethnicity, trauma and so forth round out the picture. You need to treat the patient, not just the signs and symptoms of their illness."[35]

What is special about this book is that the author is open about his own struggles and suicidality. Too often, Myers writes, doctors are "seen as above or immune to the vulnerabilities and flaws of the rest of humankind." They present and see themselves "as highly-educated professionals, solid as a rock, healers driven by an immutable code who can be counted on to take care of us when we're sick, applying their wisdom and medical or surgical prowess to make us better." Indeed, for many of us this role has become part of our personal identity—which may be useful to help our patients but may also become a problem in therapeutic work or even in our personal relationships. Earlier in this book I disclosed my own story as a father and psychiatrist who has lost his son to suicide. How come I did not recognize the suicide risk in my own son? Was I too much engulfed in my work and my academic career as a health professional? Did my son feel that he could never live up to his father's—hidden—expectations?

In Switzerland we have an emergency counseling service for medical professionals (https://remed.fmh.ch/en/). In the United States, around-the-clock help is available through the National Suicide Prevention Lifeline at 800-273-8255. A psychiatrist-run hotline specifically for physicians is available at 888-409-0141.

I have argued that the treatment of suicidality needs more than the treatment of mental disorders; it needs specific therapeutic measures. This is of course not new. Baumeister and Heatherton suggested that therapeutic interventions should focus on teaching the individual to become more effective in dealing with negative affect and to maintain a broadly meaningful perspective, that is, avoid becoming helplessly trapped so that suicide appears to be the only solution.[36] And we know that directly addressing suicidal thoughts and behaviors in cognitive behavioral therapy reduces the risk of future suicidal behaviors.[37] Early interventions are most effective. Yet it would be unrealistic to believe that after a suicidal crisis we can "cure" people from future suicidal crises. Successful treatment must enable people to intervene in the development from thoughts to plans to actions. This view is supported by the experience with ASSIP, where we make it clear to patients that after a suicide attempt the risk for suicide and suicide attempts will be increased and that this is why it is important to establish personal safety strategies (see chapter 10).

MY RECOMMENDATIONS FOR CLINICAL PRACTICE

Key Issue 1

We need models of suicide that are meaningful to patients. Treating the suicidal behavior as a form of pathology that needs to be treated (healed) is radically different from developing jointly with the patient a model of suicide as a response to personal and emotionally stressful experiences. A suicide-as-action conceptualization puts the person's own experience in the center. Once patients feel understood as individuals, motivation to engage in a therapy will increase enormously.

Key Issue 2

We need patients who are active participants in a therapeutic encounter. Patients should be seen as people who have the capacity to reflect, learn, and change behavior. The health professional's role is to facilitate collaboratively reflecting on the suicidal development and foster insight. This requires a therapeutic relationship characterized by a shared understanding of the problem between therapist and patient. When it comes to the patient's meaning of suicide in a biographical context, the patient is the expert of their story, not the therapist.

Key Issue 3

We need to relate to the person in the patient. In clinical care, nurses may often be in a better position to gain the suicidal patient's trust, in contrast to medical doctors, who tend to represent the suicide-as-illness model, with its focus on psychopathology and diagnostic evaluation. Nursing staff—and the same may be true for other non–medically trained health professions—usually find a more direct, "down-to-earth" approach to the patients. In a clinical setting, nurses often can allow more space for the patient's own contributions to treatment outcome. Specific nursing skills have been described as "establishing a therapeutic relationship, showing trust and respect, facilitating communication, getting to know the person, sharing power and responsibility, and empowering the person.[38] However, much will depend on the communication between nurses and psychiatrists, which should ideally be part of a "therapeutic milieu" in clinical care.[39]

Key Issue 4

We need clear therapy goals. We cannot "cure" people with a history of attempted suicide from suicidality—patients and therapists must be aware of the risk of future suicidal crises. The goal must be to motivate patients to collaborate with the therapist to develop and establish strategies to stay safe and to know how to deal with future existential crises. The health professional's contributions to create insight should include psychoeducation about concepts such as mental pain, the suicidal mode, dissociation and loss of control, and the role of suicide risk factors (including depression and other disorders of mental health).

Key Issue 5

We need therapy protocols that are brief, focused, and structured. This should ideally include clear within-session goals and procedures.[40]

Key Issue 6

We need to be aware that suicide risk assessment and prediction is a fallacy. We need to distinguish between long-term and short-term risk. Long-term risk assessment is based on risk factors such as a history of suicide attempts; psychiatric disorders such as depression, bipolar disorder, substance use, personality disorders, etc.; and a family history of suicidal behavior. Short-term risk can only be assessed collaboratively with the patient and should follow an empathic and narrative interviewing style fostering trust and active patient engagement. This will lead to shared decision making about the indicated procedure. What does the patient need in order to be safe for the next few hours and days? An additional factor is the clinician's "gut feeling." Feeling uneasy or recognizing conflicting emotional reactions may be signs to the interviewer that the patient is not fully participating.

Key Issue 7

We need to engage patients admitted to emergency rooms into treatment as early as possible. This requires an established procedure to refer suicidal patients to a therapy program targeting suicidal thoughts and behavior. Examples are the ED-SAFE Study, an effective combination of brief interventions administered both during and after the ER visit and the Coping Long Term with Active Suicide Program.[41] ASSIP, the brief and highly structured treatment protocol described in chapter 10, is ideally initiated as early as possible after attempted suicide. In Bern, ASSIP is not "offered" to patients as a choice but is declared as the indicated and usual clinical procedure after a suicide attempt, in addition to treatment as usual (a procedure that still gives the patient the option to refuse!). Ideally, for any early treatment engagement, the therapist contacts patients in the ER, informs them about the therapy, gives them written information about the therapy sessions and goals, and makes a first appointment. An early therapeutic relationship established while in the hospital will reduce nonattendance in follow-up.

Key Issue 8

We need to be aware of the power of a long-term therapeutic relationship with a trusted clinician. As, after attempted suicide, the risk is elevated over years, a long-term professional therapeutic contact can provide a sense of security to people, which has been beautifully expressed in this patient's statement: "He [the general practitioner] was like a rock. He really was, he was genuinely concerned for me and I could tell he was. He was really worried and in a way he made me feel better to know that someone cared and he, you know, he would see me every, maybe every month every two months just to see how everything was and till he retired really so he was a great help."[42] This is an example of a "secure anchorage" with the family doctor, in the sense of Bowlby's "secure base" concept, that is, a person we can turn to "in times of calamity," a long-term therapeutic relationship, even if appointments are spaced over months.

Key Issue 9

We need to increase the capacities for early interventions after attempted suicide. Psychological treatments, so far, play a marginal role in the provision of care for suicidal patients. A major problem is the huge discrepancy between the number of suicide attempts and the limitations of the health care systems in the provision of therapies for suicidal patients. However, hope comes from studies showing that brief and specific therapies can be highly effective. We clearly need to evaluate more therapy models in the treatment of patients at risk of suicide and in their long-term efficacy. One of the factors related to therapy adherence and outcome is the quality of the therapeutic relationship. In my opinion, future treatment studies should always include (patient-rated) measures of the therapeutic relationship, such as the Helping Alliance Questionnaire or the Working Alliance Inventory, Short Form.[43]

SUMMARY

In this chapter I have argued that the prevalent suicide-as-illness model precludes clinicians from empathically relating to the suicidal patient as a human being in crisis. To find an understanding of a person's inner experience of mental pain, personal vulnerability, and the triggering experience requires a collaborative, person-centered approach, that is, a patient who is an active participant in a meaningful therapeutic encounter. A narrative approach, in which patients are seen as individuals who have their—subjective—reasons for suicidal behavior, reduces stress in the clinician and increases patient satisfaction with the therapeutic relationship. A similar approach is recommended for suicide risk assessment. In a collaborative risk assessment, patients are invited to jointly, that is, together with the interviewer, find out what they need to be safe, in a short-term as well as long-term perspective. This stands in contrast to the traditional approach, which leaves the clinician solely responsible for categorizing a patient's suicide risk, based on psychopathological assessment and suicide risk scales. Suicidal behavior is inherently individual and personal. Risk assessment requires active patients in jointly deciding on personal safety planning.

Suicide risk factors, such as a history of suicidal behavior, and mental health problems must be included in the risk assessment. In contrast to the narrative interview, in which the patient is the expert of his or her suicide story, it is now the health professional who is the expert in the diagnostic assessment and indicated treatment. Considering the clinician's two roles, the Aeschi Working Group has recommended that interviews should always begin with the patient-centered narrative approach in order to establish a therapeutic relationship, before the active exploration of the mental state.

Appendix 1

SUICIDE IS NOT A RATIONAL ACTION: ASSIP HANDOUT FOR PATIENTS

We may have moments in our lives where we consider suicide as a possible solution to a difficult life situation. But *thinking* about suicide and *acting* on suicidal thoughts are two different things.

PSYCHOLOGICAL PAIN

A dangerous suicidal crisis is usually related to a painful emotional experience that we perceive as a threat to our existence. People often describe it as an intense psychological or mental pain. The situation becomes dangerous when we see no solution to such a painful experience.

Do you personally know such psychological conditions? Please describe your own experience.

EMOTIONAL STRESS AFFECTS BRAIN FUNCTION

Psychological pain can become so extreme that it activates our alarm system, with the release of stress hormones in our body.

High levels of stress hormones dramatically change brain function. The emotional part of the brain takes control and makes us believe that ending our life is the only way to end the pain. The brain in this moment has only one goal: to immediately end this unbearable inner experience. People who survive a suicide attempt report that suddenly they were acting like they were in a state of trance, like a robot, without feeling pain. This brain condition is called "dissociation" and happens under high emotional stress. In this mental state we are no longer able to think and decide rationally.

If you know such an experience of dissociation, can you please describe it?

EARLY STRESSFUL EXPERIENCES MAY INCREASE SUICIDE RISK

Painful experiences in our childhood can make us vulnerable for later negative experiences. For instance, abuse, maltreatment, and violence in the family but also emotional neglect and separation may increase the risk of suicide.

Is it possible that early negative experiences could be a factor in your case?

Your comments:

DEPRESSION INCREASES RISK

Depression is an important risk factor for suicide and attempted suicide. When we are feeling depressed, we tend to be ashamed and to blame ourselves. We lose hope that things will ever get better. Depressed people are often ashamed of their feelings, and they hide the pain and the shame. This makes the situation worse. Real depression requires proper treatment.

____ I believe that depression is a factor in my case
____ I may have other problems with mental health
Please describe:

ALCOHOL OR DRUG USE INCREASES RISK

Alcohol or drug use increases the risk. When we are drunk or high on drugs, suicidal thoughts may increase rapidly, the immediate situation may seem much worse, and there may be an urge to act on suicidal thoughts, without thinking.

____ I believe that alcohol or drug use is a factor in my case
Please describe:

A SUICIDAL CRISIS WILL BE STORED IN THE BRAIN

Once we have experienced a suicidal crisis, particularly after a suicide attempt, such an emotionally stressful experience will be stored in the brain. This is called the "suicidal mode." It means that the brain has learned that for unbearable mental pain suicide may indeed be a solution. This is why it is so important to develop effective safety strategies for future crises.

Your comments regarding your own past suicidal crises:

YOU SHOULD KNOW WHAT TO DO WHEN YOU ARE SUICIDAL

When we have thoughts of suicide, it is important to turn to a trusted person and talk about what occupies our mind. After a suicide attempt, it is extremely important to be aware of **early warning signs** and to know what to do *before* the suicidal mode is "**switched on**," that is, before the brain activates our alarm system and the emotional brain takes over the control. The personal safety plan includes self-help strategies like walking the dog or going to see the neighbor, but, above all, it should include the contact information of persons we can turn to in times of crisis. These may be family members or friends; it may be the emergency number of a mental health center, a family doctor, psychiatrist, or psychologist. These are professional people who can help you be safe in times of crisis.

In my case I think the following strategies could be helpful:

(1) _____
(2) _____
(3) _____
(4) _____
. . .

Trusted persons to whom I could turn:

(1) _____
(2) _____
(3) _____
(4) _____
. . .

If you are in treatment, I recommend that you print this document, add your personal comments, and take it to your therapist.

K. Michel and A. Gysin-Maillart, *Attempted Suicide Short Intervention Program ASSIP: A Manual for Clinicians* (Göttingen: Hogrefe, 2015). Adapted by K. Michel.

Appendix 2

QUESTIONNAIRE: LEARNING OBJECTIVES

(1) WHAT DOES SUICIDE AS A TOPIC MEAN TO YOU?

____ I am not personally affected
____ I know someone who died by suicide
____ I lost a dear friend by suicide
____ I lost a close family member by suicide
____ I am having/have had suicidal thoughts myself
____ I have made one or more suicide attempts
____ I am currently in a suicidal crisis

(2) DO YOU REMEMBER THE FIRST TIME YOU HEARD/LEARNED THAT SOME PEOPLE END THEIR LIVES BY SUICIDE?

____ Early years; how?
____ At school; how?
____ In teenage/adolescence; how?
____ Later; how?
____ What were your thoughts about suicide at that time?

(3) WHICH CONCEPTS COVERED IN THIS BOOK ARE PERSONALLY RELEVANT FOR YOU?

____ Depression and suicide (chapter 3)

Depression and other mental health disorders are important risk factors for suicide and need treatment. They are not the cause of suicide.

____ Psychological or mental pain (chapter 6)

Essential needs or life goals are part of our sense of identity. A negative experience can have the dimension of an assault to the sense of identity and trigger mental pain, just like pain caused by a physically life-threatening experience.

____ The suicidal mode (chapter 4)

Self-harm as a response to exceptional and threatening experiences, which, once established, will be saved as a response pattern that can be switched on again by a similar triggering event. After a first suicide attempt, the threshold for triggering the suicidal mode is lowered considerably.

____ Dissociative symptoms (chapter 4)

Typical descriptions of dissociation are frequent in the acute suicidal mode. They include feeling numb, acting in a trance or as if in a dream, functioning in the "autopilot mode," not feeling connected to one's own body, no sense of somatic pain (analgesia) and an altered sense of time.

(4) IF YOU HAVE EXPERIENCED A SUICIDAL CRISIS IN YOUR LIFE, CAN YOU IMAGINE WHAT YOUR CASE FORMULATION WOULD LOOK LIKE?

See the structure of the ASSIP case conceptualizations (chapter 10):

- What are the important basic needs and goals that characterize you as a person?
- Are there negative childhood experiences that have had a long-lasting effect on you?
- What are your vulnerable spots where you can easily be hurt?
- What events can trigger emotional stress or mental pain that may be difficult to control?
- What suicidal crises (thoughts and actions) have you experienced in your life?

(5) WHAT ARE YOUR PERSONAL WARNING SIGNS OF AN IMPENDING SUICIDAL CRISIS?

Warning signs may include changes in thinking, affect, bodily functions (stress-related symptoms), and behavior.

- Thoughts . . .
- Feelings . . .
- Changes in bodily functions . . .
- Behavior...

Some examples of what patients have told me:

> My first signs are being awake at night, nausea, and dizziness
> My thoughts go in circles
> I think that it would be a relief for me to leave this world
> My emotional life is shut down
> I don't feel myself any more
> I withdraw from people and don't want to see anyone
> I feel depressed and lose all hope that things might improve
> I think I have no right to live

I feel that I have completely messed up my life and that there is no hope that things will get better
... etc.

(6) WHAT SAFETY STRATEGIES DO YOU USE IF/WHEN THE SUICIDAL URGE INCREASES?

From our work with patients we have learned that every individual has their own preferred strategies. Who are the trusted persons you can contact at any time? Are their numbers saved in your contacts? Who are the health professionals and services you can use in case of an emergency?

Safety strategies: some examples of what patients told me:

EARLY SAFETY STRATEGIES

I go jogging, cycling (physical activity)

I cuddle with my cat

I play my clarinet

I contact someone by texting, calling, etc.; it may be helpful to chat about anything (weather, sport, politics)

I don't isolate myself, and I KEEP THE PHONE ON!

I listen to podcasts or audio books (true crime and fantasy)

I clean the windows or the bathroom

I go to the supermarket

... etc.

WHEN THE SUICIDAL URGE INCREASES

I call the family doctor or my therapist

I call the crisis line

I do *not* drink alcohol

I do *not* write a goodbye letter

I call the mental health clinic
I call the ambulance
I go directly to the emergency department
... etc.

The message is: Even seemingly unbearable emotional pain does not last forever! Mental pain will not kill you—it is your brain that panics. Don't forget: Most people who have survived a suicide attempt say that suicide was the wrong decision. But they only realize afterward. Others didn't get a second chance.

If you are a person who is directly affected by suicidal thoughts and plans, you may consider sharing your answers with a trusted person, a family member, family doctor, therapist, teacher, clergy member, or counselor. In case you are confronted with a friend or family member in an emotional crisis, asking about suicidal thoughts is never wrong but may be the means to get the person to seek lifesaving help. In appendix 3 you will find some advice about how you can address the issue.

More: https://konradmichel.com/.

Appendix 3

WHAT TO DO IF YOU ARE CONCERNED ABOUT SOMEONE

HOW TO TALK TO A PERSON IN AN EMOTIONAL CRISIS

What to say (1): "I am concerned about you. I have recently felt that you may not be well. Would you like to talk about it? Are there reasons for me to think that suicide might be an issue for you?"

What to say (2): "I care about you. I'll be there for you if you would like to talk." Suicidal people need someone who takes time to listen. Be open-minded and let yourself be surprised by whatever the person tells you. Avoid asking "why," because the immediate triggering event may be banal (for instance, a remark made by someone else) and not help with understanding the history of a person's vulnerability. Allow tears, sadness, anger, and frustration. When the person mentions suicide as a solution, say that you don't think that suicide is a good decision but that you want to find out together what could help the person be safe.

What to say (3): "Is there someone else who knows about your situation? Would you like to tell me who?" This could be a family member, friend, counselor, family doctor, etc.

OFFER TO BE AVAILABLE, BUT BE AWARE OF YOUR LIMITS

You cannot be responsible for another person when it may be a matter of life and death. If you are not a health professional, you cannot be the therapist of a family member, friend, colleague, or neighbor. But small outreach gestures like text messages, neighborly home visits, etc., can be very important for suicidal persons.

HOLD BACK WITH GIVING ADVICE

The first thoughts that cross your mind may not be helpful to the person in crisis. If you want to offer help, avoid being directive (you should . . .) but talk in the first person (listening to you makes me think that . . .). You will be most helpful if you respect the person's need for autonomy.

WHAT NOT TO SAY

"Think of your loved ones." "Think of what it would mean for your family, friends, etc." Or: "I know how you feel."

TRY TO FIND OUT THE PERSON'S ATTITUDE TOWARD GETTING PROFESSIONAL HELP

What are their past experiences? Who helped in the past? A family doctor, counselor, psychiatrist, crisis line worker, hospital admission? Has medication helped in the past? What are the barriers to seeking professional help? Try to collaboratively develop a possible plan to get help. Again: Respect the person's need for autonomy.

IF YOUR COMMITMENT CONTINUES, LEARN MORE ABOUT SUICIDE

You will find the basics in appendix 1. A history of attempted suicide increases the suicide risk. Maybe the person tells you about prior suicidal crises or about past suicide triggers. You may ask about warning signs. Consider the difference between long-term risk and the acute suicidal risk, which is usually triggered by an interpersonal conflict, acting as a trigger. The driver for the suicidal urge is usually psychological pain, which appears unbearable to the person. The important point is that people can learn to deflect the dangerous development toward a suicide action by developing personal safety strategies.

ASK WHAT HAS HELPED IN THE PAST TO CALM DOWN IN A CRISIS

It can be anything from self-soothing with a hot water bottle on the abdomen, to cuddling with the cat, to Sudoku, to outdoor activities, to connecting with others, etc.

TOGETHER WITH THE PERSON AT RISK, YOU MAY EXPLORE ONLINE SUPPORT

Offer help in tackling the mental health care system and in how to make an appointment. You may offer to accompany the person to a first appointment (see appendix 4).

IF YOUR INVOLVEMENT CONTINUES, SEEK PROFESSIONAL ADVICE FOR YOURSELF

You can call a crisis line or contact a mental health center without disclosing the person's name. Be open to the suicidal person about this. Tell the person what advice you received, where to turn to, and why it is necessary to get a professional mental health assessment. Again: when the person's suicidal condition is serious, the situation requires professional involvement. You cannot be responsible for another person's life.

Appendix 4

RESOURCES

CRISIS LINES: USA

National Suicide Prevention Lifeline: 800-273-TALK (8255)
Samaritans Helpline: 877-870-4673 (HOPE): call or text
The Trevor Project: 866-488-7386: a hotline for LGBT youth

CRISIS LINES: UK

Papyrus: Prevention of Young Suicide, https://www.papyrus-uk.org/. A national charity dedicated to the prevention of young suicide. It operates HOPELINEUK, which offers support and advice to young people. Phone: 0800 068 41 41, every day, 9 a.m. to midnight. Text: 0786 003 9967.

CRISIS LINES: NEW ZEALAND

Lifeline: https://www.lifeline.org.nz/services/suicide-crisis-helpline.

CRISIS LINES: AUSTRALIA

Lifeline (chat): https://www.lifeline.org.au/crisis-chat/.
Lifeline (text): https://www.lifeline.org.au/crisis-text/.
Lifeline (general information): https://www.lifeline.org.au/.

CRISIS LINES: SWITZERLAND

Tel 143: https://www.143.ch/.
Pro Juventute: https://www.147.ch/. For children and adolescents.

CRISIS LINES: INTERNATIONAL

Suicide.org, includes International Suicide Hotlines: http://www.suicide.org/index.html.
Crisis text lines: http://www.crisistextline.org. 741741 (U.S.); 686868 (Canada); 85258 (UK), Ireland 50808. Connect with a trained crisis counselor 24/7.
National Suicide Crisis Lines: https://en.wikipedia.org/wiki/List_of_suicide_crisis_lines.
Befrienders Worldwide: http://www.befrienders.org. An international network to provide emotional support services for people who are suicidal and/or in distress.

ONLINE HELP

Suicide Prevention Resource Guide: https://www.healthline.com/health/mental-health/suicide-resource-guide.
American Foundation for Suicide Prevention: https://afsp.org/get-help.
Suicide Ireland: http://www.suicideireland.com/.
Help Guide: An impressive resource for getting help, information, education, videos, and much more. For suicide: https://www.helpguide.org

/articles/suicide-prevention/are-you-feeling-suicidal.htm; for depression: https://www.helpguide.org/home-pages/depression.htm.

Beyond Blue: https://www.beyondblue.org.au/the-facts/suicide-prevention. There are lots of resources for people considering suicide. https://www.beyondblue.org.au/get-support/get-immediate-support gives information on how to reach people who would usually not get in touch with a medical or mental health service.

JT1 (Just Tell One): http://justtellone.org/. Offers support for depression, alcohol, drugs, and suicide.

Support Line: https://www.supportline.org.uk/problems/suicide/. Includes many helpful links.

The Trevor Project: https://www.thetrevorproject.org/. Offers information for suicide prevention in lesbian, gay, bisexual, and transgender youth.

Staying safe from suicidal thoughts: https://www.stayingsafe.net/home. Includes a range of mental health programs including safety planning resources.

Stanley-Brown Safety Plan: https://apps.apple.com/us/app/safety-plan/id695122998.

Virtual Hope Box: https://apps.apple.com/ch/app/virtual-hope-box/id825099621. A well-designed smartphone application that contains a large number of tools to help users cope with stressful situations. The tool that appealed to me most is the "Controlled Breathing" exercise.

Save: https://save.org/. Founded by a group of survivors, with the mission to prevent suicide through public awareness and education and to serve as a resource to those touched by suicide.

Ipsilon: https://www.ipsilon.ch/. An official Swiss site with helpful links.

Pro Juventute Switzerland: https://www.147.ch/. Lots of information, online chat and crisis line.

SUPPORT FOR PEOPLE BEREFT BY SUICIDE

Alliance of Hope: https://allianceofhope.org/. The site includes a section for professionals.

Help Guide: https://www.helpguide.org/articles/grief/coping-with-a-loved-ones-suicide.htm.

American Association of Suicidiology: https://suicidology.org/resources/suicide-loss-survivors/. Includes *The SOS Handbook*, a quick-reference booklet for suicide survivors

Suicidebereavement UK: https://suicidebereavementuk.com/.

Suicide (SOBS): http://www.uk-sobs.org.uk. A national charity providing dedicated support to adults who have been bereaved by suicide.

SUICIDE PREVENTION: GENERAL

International Association for Suicide Prevention (IASP): https://www.iasp.info/. Hosts a database of organizations that provide crisis support in every continent.

SUICIDE PREVENTION: USA

Suicide Prevention Resource Center: https://www.sprc.org/.
American Association of Suicidology: https://suicidology.org/.
Zero Suicide: https://zerosuicide.edc.org/.
QRT Institute: http://www.qprinstitute.com.
Online Therapist Finder: https://www.onlinetherapy.com/suicide-prevention/.
Suicide Prevention Therapist Finder: www.helppro.com.

SUICIDE PREVENTION: IRELAND

National Office for Suicide Prevention: https://www.hse.ie/eng/services/list/4/mental-health-services/nosp/.

National Suicide Research Foundation https://www.nsrf.ie/.

SUICIDE PREVENTION: UK

Zero Suicide Alliance https://www.zerosuicidealliance.com/.
National Suicide Prevention Alliance (NSPA): https://www.nspa.org.uk.

SUICIDE PREVENTION: AUSTRALIA

https://www.suicidepreventionaust.org/.

OTHER RESOURCES

Recommendations for suicide reporting in media: https://reportingonsuicide.org/recommendations/.
Suicide and guns: https://www.bradyunited.org/report/suicide-prevention-report-2018.
Livingworks: ASIST https://www.livingworks.net/asist.
SafeTALK: http://www.livingworks.net/programs/safetalk/.

BOOKS

Stacey Freedenthal, *Helping the Suicidal Person* (Routledge, 2017). A book for a wide range of health professionals who in their daily work may be faced with suicide as an issue.

Rory O'Connor, *When It Is Darkest: Why People Die by Suicide and What We Can Do to Prevent It* (Vermilion, 2022). An evidence-based self-help book on understanding and preventing suicide.

Susan Rose Blauner, *How I Stayed Alive When My Brain Was Trying to Kill Me* (HarperCollins, 2009). An exceptional book, with strong personal messages from the author's own impressive story of severe suicidality and long-term therapy. It contains such important messages as the "the brain has a mind of its own" and "it's okay to have suicidal thoughts, just don't act on them."

Thomas Joiner, *Why People Die by Suicide* (Harvard University Press, 2007). The author's personal experience of losing his father by suicide. A new theoretical model of suicide is introduced in a refreshing, non-academic way.

P. Omerov and J. Bullington, "Nursing Care of the Suicidal Patient," in *Suicide Risk Assessment and Prevention*, ed. M. Pompili (Springer, 2021). Highly recommended for nursing staff (https://doi.org/10.1007/978-3-030-41319-4_65-1).

NOTES

INTRODUCTION

1. L. Valach, R. A. Young, and K. Michel, "Understanding Suicide as an Action," in *Building a Therapeutic Alliance with the Suicidal Patient*, ed. K. Michel and D. A. Jobes (Washington, DC: American Psychological Association, 2011).
2. K. Michel and L. Valach, "Suicide as Goal-Directed Action," *Archives of Suicide Research* 3, no. 3 (1997).
3. K. Michel et al., "Discovering the Truth in Attempted Suicide," Case Reports, *American Journal of Psychotherapy* 56, no. 3 (2002), http://www.ncbi.nlm.nih.gov/pubmed/12400207.
4. J. R. Rogers and K. M. Soyka, "'One Size Fits All': An Existential-Constructivist Perspective on the Crisis Intervention Approach with Suicidal Individuals," *Journal of Contemporary Psychotherapy* 34, no. 1 (2004).
5. H. Ritchie, M. Roser, and E. Ortiz-Ospina, "Suicide," OurWorldInData.org, 2015, https://ourworldindata.org/suicide.
6. S. C. Curtin and M. F. Garnett, "Provisional Numbers and Rates of Suicide by Month and Demographic Characteristics: United States, 2021," Vital Statistics Rapid Release no. 24, September 2022, https://dx.doi.org/10.15620/cdc:120830.
7. CDC, "Preventing Suicide," October 2022, https://www.cdc.gov/suicide/pdf/NCIPC-Suicide-FactSheet-508_FINAL.pdf.

I. LOSING A PATIENT TO SUICIDE AND WHAT IT MEANS FOR A YOUNG DOCTOR

1. O. Grad and K. Michel, "Therapists as Client Suicide Survivors," in *Therapeutic and Legal Issues for Therapists Who Have Survived a Client Suicide: Breaking the Silence*, ed. K. M. Weiner (New York: Haworth, 2005).
2. J. A. Motto, "The Impact of Patient Suicide on the Therapist's Feelings," *Weekly Psychiatry Update Series* 3, no. 21 (1979).
3. J. Morrissey and A. Higgins, "'When My Worst Fear Happened': Mental Health Nurses' Responses to the Death of a Client Through Suicide," *Journal of Psychiatric Mental Health Nursing* 28, no. 5 (October 2021), https://doi.org/10.1111/jpm.12765.
4. A. E. Hohler, "Selbsttötung als Panne" (Self-murder as mishap), letter to the editor, *Tages Anzeiger Magazin* (Zürich), February 13, 1988.
5. B. M. Barraclough and D. J. Pallis, "Depression Followed by Suicide: A Comparison of Depressed Suicides with Living Depressives," *Psychological Medicine* 5, no. 1 (February 1975), https://doi.org/10.1017/s0033291700007212.
6. G. E. Murphy, "The Physician's Responsibility for Suicide, II: Errors of Omission," *Annals of Internal Medicine* 82, no. 3 (March 1975), https://doi.org/10.7326/0003-4819-82-3-305.
7. G. E. Murphy, "The Physician's Responsibility for Suicide, I: An Error of Commission," *Annals of Internal Medicine* 82, no. 3 (March 1975), https://doi.org/10.7326/0003-4819-82-3-301.
8. Thomas A. M. Kramer, "Endogenous Versus Exogenous: Still Not the Issue," *Medscape Psychopharmacology Today* (2002), https://www.medscape.com/viewarticle/418269; C. A. Watts, "The Evolution of Depressive Symptoms in Endogenous Depression," *Journal of the Royal College of General Practitioners* 15, no. 4 (April 1968), https://www.ncbi.nlm.nih.gov/pubmed/5653294.
9. J. Fawcett, "Suicidal Depression and Physical Illness," *JAMA* 219, no. 10 (March 1972), https://www.ncbi.nlm.nih.gov/pubmed/5066772.
10. H. Hafner, "Descriptive Psychopathology, Phenomenology, and the Legacy of Karl Jaspers," *Dialogues in Clinical Neuroscience* 17, no. 1 (March 2015), https://www.ncbi.nlm.nih.gov/pubmed/25987860.

11. Karl Jaspers, *General Psychopathology*, trans. J. Hoenig and M. Hamilton (Manchester: Manchester University Press, 1963), 19.
12. A. B. Smulevich, "Sluggish Schizophrenia in the Modern Classification of Mental Illness," *Schizophrenia Bulletin* 15, no. 4 (1989), https://doi.org/10.1093/schbul/15.4.533.
13. K. Michel, "Könnte der Arzt mehr tun?" (Could the medical profession do more?), *Schweizerische Medizinische Wochenschrift/Journal Suisse de Médecine* 116 (1986).
14. K. Michel, "Suicide Risk Factors: A Comparison of Suicide Attempters with Suicide Completers," *British Journal of Psychiatry* 150 (January 1987), https://doi.org/10.1192/bjp.150.1.78.
15. K. Michel, "Suicide in Young People Is Different," *Crisis* 9, no. 2 (November 1988), https://www.ncbi.nlm.nih.gov/pubmed/3215036.

2. FIRST LESSONS IN REDUCING SUICIDE

1. W. Rutz et al., "An Educational Program on Depressive Disorders for General Practitioners on Gotland: Background and Evaluation," *Acta Psychiatrica Scandinavica* 79, no. 1 (1989).
2. E. T. Isometsä et al., "The Last Appointment Before Suicide: Is Suicide Intent Communicated?," *American Journal of Psychiatry* 152, no. 6 (1995).
3. K. Michel and L. Valach, "Suicide Prevention: Spreading the Gospel to General Practitioners," *British Journal of Psychiatry* 160 (June 1992), https://doi.org/10.1192/bjp.160.6.757.
4. Monica M. Matthieu et al., "Evaluation of Gatekeeper Training for Suicide Prevention in Veterans," *Archives of Suicide Research* 12, no. 2 (2008).
5. "QPR Gatekeeper Training for Suicide Prevention: The Model, Theory and Research," 2013, https://www.qprinstitute.com/uploads/QPR%20Theory%20Paper.pdf.
6. Swiss Medical Association, *Krise und Suizid, Basisdokument* (Bern, 1992), https://www.fmh.ch/files/pdf14/Suizid_Basisdokument_D.pdf.
7. G. Caplan, *Principles of Preventive Psychiatry* (New York: Basic Books, 1964).

8. V. Ajdacic-Gross et al., "In-patient Suicide—a 13-Year Assessment," *Acta Psychiatrica Scandinavica* 120, no. 1 (July 2009), https://doi.org/10.1111/j.1600-0447.2009.01380.x.
9. CDC, "Suicide Contagion and the Reporting of Suicide: Recommendations from a National Workshop," Recommendations and Reports 1994, https://www.cdc.gov/mmwr/preview/mmwrhtml/00031539.htm.
10. K. Michel et al., "An Exercise in Improving Suicide Reporting in Print Media," *Crisis* 21, no. 2 (2000).
11. G. Sonneck, E. Etzersdorfer, and S. Nagel-Kuess, "Imitative Suicide on the Viennese Subway," *Social Science & Medicine* 38, no. 3 (1994).
12. A. Schmidtke et al., "Attempted Suicide in Europe: Rates, Trends and Sociodemographic Characteristics of Suicide Attempters During the Period 1989–1992. Results of the WHO/EURO Multicentre Study on Parasuicide," *Acta Psychiatrica Scandinavica* 93, no. 5 (May 1996), https://doi.org/10.1111/j.1600-0447.1996.tb10656.x.
13. U. Bille-Brahe et al., "A Repetition-Prediction Study of European Parasuicide Populations: A Summary of the First Report from Part II of the WHO/EURO Multicentre Study on Parasuicide in Co-operation with the EC Concerted Action on Attempted Suicide," Comparative Study Multicenter Study, *Acta Psychiatrica Scandinavica* 95, no. 2 (February 1997), http://www.ncbi.nlm.nih.gov/pubmed/9065670.
14. J. Bancroft et al., "The Reasons People Give for Taking Overdoses: A Further Inquiry," *British Journal of Medical Psychology* 52, no. 4 (December 1979), https://doi.org/10.1111/j.2044-8341.1979.tb02536.x.
15. K. Michel, L. Valach, and V. Waeber, "Understanding Deliberate Self-Harm: The Patients' Views," *Crisis* 15, no. 4 (1994), http://www.ncbi.nlm.nih.gov/pubmed/7729149.
16. E. S. Shneidman, *The Definition of Suicide: An Essay* (New York: John Wiley & Sons, 1985).
17. Michel, Valach, and Waeber, "Understanding Deliberate Self-Harm."

3. EMOTIONAL STRESS AFFECTS BRAIN FUNCTION

1. M. Iacoboni et al., "Grasping the Intentions of Others with One's Own Mirror Neuron System," *PLoS Biology* 3, no. 3 (March 2005), https://doi.org/10.1371/journal.pbio.0030079.

2. M. D. Rudd, "The Suicidal Mode: A Cognitive-Behavioral Model of Suicidality," *Suicide and Life-Threatening Behavior* 30, no. 1 (2000).
3. J. D. Bremner et al., "Neural Correlates of Memories of Childhood Sexual Abuse in Women with and Without Posttraumatic Stress Disorder," *American Journal of Psychiatry* 156, no. 11 (1999).
4. Ruth A. Lanius et al., "Brain Activation During Script-Driven Imagery Induced Dissociative Responses in PTSD: A Functional Magnetic Resonance Imaging Investigation," *Biological Psychiatry* 52, no. 4 (2002).
5. T. Reisch et al., "An fMRI Study on Mental Pain and Suicidal Behavior," *Journal of Affective Disorders* 126, no. 1 (2010).
6. M. G. Griffin, P. A. Resick, and M. B. Mechanic, "Objective Assessment of Peritraumatic Dissociation: Psychophysiological Indicators," *American Journal of Psychiatry* 154, no. 8 (August 1997), https://doi.org/10.1176/ajp.154.8.1081.
7. C. R. Marmar et al., "Peritraumatic Dissociation and Posttraumatic Stress in Male Vietnam Theater Veterans," *American Journal of Psychiatry* 151, no. 6 (June 1994), https://doi.org/10.1176/ajp.151.6.902.
8. D. Spiegel and E. Cardena, "Disintegrated Experience: The Dissociative Disorders Revisited," *Journal of Abnormal Psychology* 100, no. 3 (August 1991), https://doi.org/10.1037//0021-843x.100.3.366.
9. Rudd, "The Suicidal Mode: A Cognitive-Behavioral Model of Suicidality."
10. M. D. Rudd and G. K. Brown, "A Cognitive Theory of Suicide: Building Hope in Treatment and Strengthening the Therapeutic Relationship," in *Building a Therapeutic Alliance with the Suicidal Patient*, ed. K. Michel and D. A. Jobes (Washington, DC: APA, 2011).
11. American Psychiatric Association, *The Diagnostic and Statistical Manual of Mental Disorders*, 5th ed. (Washington, DC.: American Psychiatric Publishing, 2013).
12. I. Orbach, "Dissociation, Physical Pain, and Suicide: A Hypothesis," review, *Suicide and Life-Threatening Behavior* 24, no. 1 (Spring 1994), http://www.ncbi.nlm.nih.gov/pubmed/8203010.
13. I. Weinberg, "The Prisoners of Despair: Right Hemisphere Deficiency and Suicide," *Neuroscience and Biobehavioral Reviews* 24, no. 8 (December 2000), https://doi.org/10.1016/s0149-7634(00)00038-5.
14. "Kevin Hines, Survivor, Storyteller, Film Maker," https://www.kevinhinesstory.com/.

4. THE BRAIN AND SUICIDE

1. E. R. Kandel, "A New Intellectual Framework for Psychiatry," *American Journal of Psychiatry* 155, no. 4 (April 1998), https://doi.org/10.1176/ajp.155.4.457.
2. E. R. Kandel, "Psychotherapy and the Single Synapse. The Impact of Psychiatric Thought on Neurobiologic Research," *New England Journal of Medicine* 301, no. 19 (November 8, 1979), https://doi.org/10.1056/NEJM197911083011904.
3. P. Seeger, *In Search of Memory: The Neuroscientist Eric Kandel*, 2009, https://icarusfilms.com/if-mem.
4. B. Libet et al., "Time of Conscious Intention to Act in Relation to Onset of Cerebral Activity (Readiness-Potential). The Unconscious Initiation of a Freely Voluntary Act," *Brain* 106 (part 3) (September 1983), https://doi.org/10.1093/brain/106.3.623.
5. D. M. Wegner, *The Illusion of Conscious Will* (Cambridge, MA: MIT Press, 2002).
6. J. Bruner, *Making Stories: Law, Literature, Life* (Cambridge, MA: Harvard University Press, 2002).
7. M. Asberg, L. Traskman, and P. Thoren, "5-HIAA in the Cerebrospinal Fluid. A Biochemical Suicide Predictor?," *Archives of General Psychiatry* 33, no. 10 (October 1976), https://doi.org/10.1001/archpsyc.1976.01770100055005.
8. J. J. Mann et al., "Relationship Between Central and Peripheral Serotonin Indexes in Depressed and Suicidal Psychiatric Inpatients," *Archives of General Psychiatry* 49, no. 6 (June 1992), http://www.ncbi.nlm.nih.gov/pubmed/1376106.
9. A. Caspi et al., "Influence of Life Stress on Depression: Moderation by a Polymorphism in the 5-HTT Gene," *Science* 301, no. 5631 (2003).
10. A. Y. Dombrovski et al., "Reward Signals, Attempted Suicide, and Impulsivity in Late-Life Depression," *JAMA Psychiatry* 70, no. 10 (October 2013), https://doi.org/10.1001/jamapsychiatry.2013.75.
11. F. Jollant et al., "Impaired Decision Making in Suicide Attempters," *American Journal of Psychiatry* 162, no. 2 (2005).
12. K. N. Ochsner et al., "Rethinking Feelings: An FMRI Study of the Cognitive Regulation of Emotion," *Journal of Cognitive Neuroscience* 14,

no. 8 (November 15, 2002), https://doi.org/10.1162/089892902760807212.

13. E. T. C. Lippard et al., "Preliminary Examination of Gray and White Matter Structure and Longitudinal Structural Changes in Frontal Systems Associated with Future Suicide Attempts in Adolescents and Young Adults with Mood Disorders," *Journal of Affect Disorders* 245 (February 15, 2019), https://doi.org/10.1016/j.jad.2018.11.097.
14. E. Isometsa et al., "Inadequate Dosaging in General Practice of Tricyclic vs. Other Antidepressants for Depression," *Acta Psychiatrica Scandinavica* 98, no. 6 (December 1998), https://doi.org/10.1111/j.1600-0447.1998.tb10118.x.
15. G. Isacsson et al., "The Utilization of Antidepressants—a Key Issue in the Prevention of Suicide: An Analysis of 5281 Suicides in Sweden During the Period 1992–1994," *Acta Psychiatrica Scandinavica* 96, no. 2 (August 1997), https://doi.org/10.1111/j.1600-0447.1997.tb09912.x.
16. G. Isacsson, "Depression is the Core of Suicidality—Its Treatment Is the Cure," *Acta Psychiatrica Scandinavica* 114, no. 3 (September 2006), https://doi.org/10.1111/j.1600-0447.2006.00871.x.
17. G. Isacsson, "Suicide Prevention—a Medical Breakthrough?," *Acta Psychiatrica Scandinavica* 102, no. 2 (August 2000), http://www.ncbi.nlm.nih.gov/pubmed/10937783.
18. H. M. Van Praag, "Why Has the Antidepressant Era Not Shown a Significant Drop in Suicide Rates?," *Crisis* 23, no. 2 (2002), https://doi.org/10.1027//0227-5910.23.2.77.
19. H. M. Van Praag, "A Stubborn Behaviour: The Failure of Antidepressants to Reduce Suicide Rates," *World Journal of Biological Psychiatry* 4, no. 4 (October 2003), http://www.ncbi.nlm.nih.gov/pubmed/14608590.
20. D. De Leo, "Suicide Prevention Is Far More Than a Psychiatric Business," *World Psychiatry* 3, no. 3 (October 2004), http://www.ncbi.nlm.nih.gov/pubmed/16633482.
21. K. Michel et al., "Psychopharmacological Treatment Is Not Associated with Reduced Suicide Ideation and Reattempts in an Observational Follow-up Study of Suicide Attempters," *Journal of Psychiatric Research* 140 (August 2021), https://doi.org/10.1016/j.jpsychires.2021.05.068.

22. R. D. Gibbons et al., "Early Evidence on the Effects of Regulators' Suicidality Warnings on SSRI Prescriptions and Suicide in Children and Adolescents," *American Journal of Psychiatry* 164, no. 9 (2007).
23. J. M. Bertolote et al., "Suicide and Mental Disorders: Do We Know Enough?," *British Journal of Psychiatry* 183 (November 2003), http://www.ncbi.nlm.nih.gov/pubmed/14594911.
24. NHS Foundation Trust, "Prescribing Guidelines for the Management of Depression," 2021, https://www.candi.nhs.uk/sites/default/files/Depression%20%20Prescribing%20Guidelines%202021.pdf.
25. I. Elkin et al., "National Institute of Mental Health Treatment of Depression Collaborative Research Program. General Effectiveness of Treatments," *Archives of General Psychiatry* 46, no. 11 (November 1989), https://doi.org/10.1001/archpsyc.1989.01810110013002.
26. J. L. Krupnick et al., "The Role of the Therapeutic Alliance in Psychotherapy and Pharmacotherapy Outcome: Findings in the National Institute of Mental Health Treatment of Depression Collaborative Research Program," *Journal of Consulting and Clinical Psychology* 64, no. 3 (1996).
27. F. Barjasteh-Askari et al., "Relationship Between Suicide Mortality and Lithium in Drinking Water: A Systematic Review and Meta-analysis," *Journal of Affect Disorders* 264 (March 1, 2020), https://doi.org/10.1016/j.jad.2019.12.027.
28. C. Andrade, "Ketamine for Depression, 6: Effects on Suicidal Ideation and Possible Use as Crisis Intervention in Patients at Suicide Risk," *Journal of Clinical Psychiatry* 79, no. 2 (March/April 2018), https://doi.org/10.4088/JCP.18f12242.

5. PROBLEMS OF COMMUNICATION IN MEDICAL CONSULTATION

1. D. Philipps, "In Unit Stalked by Suicide, Veterans Try to Save One Another," *New York Times*, September 19, 2015, https://www.nytimes.com/2015/09/20/us/marine-battalion-veterans-scarred-by-suicides-turn-to-one-another-for-help.html.
2. E. Bromet et al., "Cross-national Epidemiology of *DSM-IV* Major Depressive Episode," *BMC Medicine* 9 (July 26, 2011), https://doi.org/10.1186/1741-7015-9-90.

3. W. Rutz et al., "Prevention of Male Suicides: Lessons from Gotland Study," *Lancet* 345, no. 8948 (February 25, 1995), http://www.ncbi.nlm.nih.gov/pubmed/7861901.
4. S. J. Kuramoto-Crawford, B. Han, and R. T. McKeon, "Self-Reported Reasons for Not Receiving Mental Health Treatment in Adults with Serious Suicidal Thoughts," *Journal of Clinical Psychiatry* 78, no. 6 (June 2017), https://doi.org/10.4088/JCP.16m10989.
5. A. J. Treolar and T. J. Pinfold, "Deliberate Self-Harm: An Assessment of Patients' Attitudes to the Care They Receive," *Crisis* 14 (1993).
6. H. Gilburt, D. Rose, and M. Slade, "The Importance of Relationships in Mental Health Care: A Qualitative Study of Service Users' Experiences of Psychiatric Hospital Admission in the UK," *BMC Health Services Research* 8 (April 25, 2008), https://doi.org/10.1186/1472-6963-8-92.
7. M. Balint, "The Doctor, His Patient, and the Illness," *Lancet* 268, no. 6866 (April 2, 1955), https://doi.org/10.1016/s0140-6736(55)91061-8.
8. J. R. Rogers and K. M. Soyka, "'One size fits all': An Existential-Constructivist Perspective on the Crisis Intervention Approach with Suicidal Individuals," *Journal of Contemporary Psychotherapy* 34, no. 1 (2004).

BOX: THEORIES TO EXPLAIN SUICIDE

1. E. S. Shneidman and N. L. Farberow, *Clues to Suicide* (New York: McGraw-Hill, 1957).
2. E. S. Shneidman, "Commentary: Suicide as Psychache," *Journal of Nervous and Mental Disease* 181, no. 3 (1993).
3. I. Orbach, "Dissociation, Physical Pain, and Suicide: A Hypothesis," review, *Suicide and Life-Threatening Behavior* 24, no. 1 (Spring 1994), http://www.ncbi.nlm.nih.gov/pubmed/8203010.
4. R. F. Baumeister, "Suicide as Escape from Self," *Psychological Review* 97, no. 1 (1990): 93.
5. M. Williams, *Suicide and Attempted Suicide*, 2nd ed. (London: Penguin, 2002).
6. J. M. G. Williams and L. R. Pollock, "Psychological Aspects of the Suicidal Process," in *Understanding Suicidal Behaviour: The Suicidal*

Process Approach to Research, Treatment, and Prevention, ed. van Heeringen K. (Chichester: John Wiley & Sons, 2001).
7. Ronald W. Maris and Bernard Melvin Lazerwitz, *Pathways to Suicide: A Survey of Self-Destructive Behaviors* (Baltimore: Johns Hopkins University Press, 1981).
8. M. D. Rudd, "The Suicidal Mode: A Cognitive-Behavioral Model of Suicidality," *Suicide and Life-Threatening Behavior* 30, no. 1 (2000).
9. M. D. Rudd, T. Joiner, and M. H. Rajab, *Treating Suicidal Behavior: An Effective, Time-Limited Approach* (New York: Guilford, 2001).
10. M. M. Linehan, *Cognitive Behavioural Therapy of Borderline Personality Disorder* (New York: Guilford, 1993).
11. T. E. Joiner Jr., et al., *The Interpersonal Theory of Suicide: Guidance for Working with Suicidal Clients* (Washington, DC: American Psychological Association, 2009).
12. Kimberly A. Van Orden et al., "The Interpersonal Theory of Suicide," *Psychological Review* 117, no. 2 (2010).
13. J. C. Wakefield, "Klerman's 'Credo' Reconsidered: Neo-Kraepelinianism, Spitzer's Views, and What We Can Learn from the Past," *World Psychiatry* 21, no. 1 (February 2022), https://doi.org/10.1002/wps.20942.
14. E. Robins, *The Final Months: A Study of the Lives of 134 Persons Who Committed Suicide* (New York: Oxford University Press, 1981); Yeates Conwell, "Relationships of Age and Axis I Diagnoses in Victims of Completed Suicide: A Psychological Autopsy Study," *American Journal of Psychiatry* 153 (1996); E. C. Harris and B. Barraclough, "Suicide as an Outcome for Mental Disorders. A Meta-analysis," *British Journal of Psychiatry* 170 (March 1997), https://doi.org/10.1192/bjp.170.3.205.
15. Jonathan T. O. Cavanagh et al., "Psychological Autopsy Studies of Suicide: A Systematic Review," *Psychological Medicine* 33, no. 3 (2003).
16. J. John Mann et al., "Toward a Clinical Model of Suicidal Behavior in Psychiatric Patients," *American Journal of Psychiatry* 156, no. 2 (1999).
17. J. John Mann, "A Current Perspective of Suicide and Attempted Suicide," *Annals of Internal Medicine* 136, no. 4 (2002).
18. Thomas Insel, "Twenty Billion Fails to 'Move the Needle' on Mental Illness," interview by G. Henriques, May 23, 2017, https://www

.psychologytoday.com/us/blog/theory-knowledge/201705/twenty-billion-fails-move-the-needle-mental-illness.

6. SUICIDE IS NOT AN ILLNESS

1. S. Harding, D. Phillips, and M. Fogarty, *Contrasting Values in Western Europe* (London: MacMillan, 1986).
2. K. Michel, "General Problems in Planning and Implementing Prevention and/or Intervention Programmes," in *Proceedings from the Fourth European Symposium on Suicidal Behaviour*, ed. U. Bille-Brahe and H. Schiodt (Odense: Odense University Press, 1994).
3. K. Michel and L. Valach, "Kenntnisse und Einstellungen praktizierender Aerzte zum Thema Suizid," *Schweiz Rundschau Med (Praxis)* 79 (1990).
4. S. J. Kuramoto-Crawford, B. Han, and R. T. McKeon, "Self-Reported Reasons for Not Receiving Mental Health Treatment in Adults with Serious Suicidal Thoughts," *Journal of Clinical Psychiatry* 78, no. 6 (June 2017), https://doi.org/10.4088/JCP.16m10989.
5. J. Angst and W. Wicki, "The Zurich Study. XI. Is Dysthymia a Separate Form of Depression? Results of the Zurich Cohort Study," *European Archives of Psychiatry and Clinical Neuroscience* 240, no. 6 (1991), https://doi.org/10.1007/BF02279765.
6. K. Smith and S. Crawford, "Suicidal Behavior Among 'Normal' High School Students," *Suicide and Life Threatening Behavior* 16, no. 3 (Fall 1986), https://doi.org/10.1111/j.1943-278x.1986.tb01013.x.
7. S. Stefan, *Rational Suicide, Irrational Laws* (New York: Oxford University Press, 2016).
8. Ronald W. Maris and Bernard Melvin Lazerwitz, *Pathways to Suicide: A Survey of Self-Destructive Behaviors* (Baltimore: Johns Hopkins University Press, 1981), xvii.
9. K. Michel and L. Valach, "Suicide as Goal-Directed Action," *Archives of Suicide Research* 3, no. 3 (1997).
10. Maris and Lazerwitz, *Pathways to suicide*, 315.
11. H. M. Adler, "The History of the Present Illness as Treatment: Who's Listening, and Why Does It Matter?," *Journal of the American Board of Family Practice* 10, no. 1 (1997).

12. John Bowlby, *A Secure Base: Parent-Child Attachment and Healthy Human Development* (New York: Basic Books, 2008).
13. John Bowlby, *A Secure Base: Clinical Applications of Attachment Theory* (London: Routledge, 1988).
14. J. Holmes, *The Search for the Secure Base: Attachment Theory and Psychotherapy* (Hove: Brunner Routledge, 2001), 87.
15. L. E. Angus and J. McLeod, "Toward an Integrative Framework for Understanding the Role of Narrative in the Psychotherapy Process," in *The Handbook of Narrative and Psychotherapy, Practice, Theory, and Research*, ed. Lynne E. Angus and John McLeod (Thousand Oaks, CA: Sage, 2004).
16. J. Bruner, "The Narrative Creation of Self," in *The Handbook of Narrative and Psychotherapy: Practice, Theory, and Research*, ed. L. E. Angus and J. Mcleod (Thousand Oaks, CA: Sage, 2004), 13.
17. L. Gaston et al., "Alliance, Technique, and Their Interactions in Predicting Outcome of Behavioural, Cognitive, and Brief Dynamic Therapy," *Psychotherapy Research* 8 (1998).
18. D. Lizardi and B. Stanley, "Treatment Engagement: A Neglected Aspect in the Psychiatric Care of Suicidal Patients," *Psychiatric Services* 61, no. 12 (2010).
19. J. van Os and D. Kamp, "Putting the Psychotherapy Spotlight Back on the Self-Reflecting Actors Who Make It Work," *World Psychiatry* 18, no. 3 (October 2019), https://doi.org/10.1002/wps.20667.

7. THE FRAGILE SENSE OF WHO WE ARE

1. R. F. Baumeister, "Suicide as Escape from Self," *Psychological Review* 97, no. 1 (1990).
2. E. H. Erikson, *Identity and the Life Cycle* (New York: Norton, 1980).
3. M. M. Kahn and E. U. Syed, "Suicide in Asia: Epidemiology, Risk Factors, and Prevention," in *International Handbook of Suicide Prevention: Research, Policy, and Practice*, ed. R. C. O'Connor, S. Platt, and J. Gordon (Chichester: John Wiley & Sons, 2011).
4. B. C. Park, J. Soo Im, and K. Strother Ratcliff, "Rising Youth Suicide and the Changing Cultural Context in South Korea," *Crisis* 35, no. 2 (2014), https://doi.org/10.1027/0227-5910/a000237.

5. T. A. Carleton, "Crop-Damaging Temperatures Increase Suicide Rates in India," *Proceedings of the National Academy of Sciences USA* 114, no. 33 (August 15, 2017), https://doi.org/10.1073/pnas.1701354114.
6. G. Kaur, "The Country Where Thirty Farmers Die Each Day," CNN, March 17, 2022, https://edition.cnn.com/2022/03/17/opinions/india-farmer-suicide-agriculture-reform-kaur/index.html.
7. D. Lederer, "Honfibu: Nationhood, Manhood, and the Culture of Self-Sacrifice in Hungary," in *From Sin to Insanity: Suicide in Early Modern Europe*, ed. Jeffrey R. Watt (Ithaca, NY: Cornell University Press, 2004).
8. M. L. Rasmussen, H. Haavind, and G. Dieserud, "Young Men, Masculinities, and Suicide," *Archives of Suicide Research* 22, no. 2 (April–June 2018), https://doi.org/10.1080/13811118.2017.1340855.
9. E. L. Meerwijk and S. J Weiss, "Toward a Unifying Definition of Psychological Pain," *Journal of Loss and Trauma* 16, no. 5 (2011), https://doi.org/10.1080/15325024.2011.572044.
10. E. S. Shneidman, "Commentary: Suicide as Psychache," *Journal of Nervous and Mental Disease* 181, no. 3 (1993).
11. E. Bolger, "Grounded Theory Analysis of Emotional Pain," *Psychotherapy Research* 9, no. 3 (1999), https://doi.org/10.1080/10503309912331332801.
12. C. A. Soper, *The Evolution of Suicide*, ed. Todd K. Shackelford and Viviana A. Weekes-Shackelford (Springer International Publishing AG, 2018), 251.
13. R. Bruffaerts et al., "Childhood Adversities as Risk Factors for Onset and Persistence of Suicidal Behaviour," *British Journal of Psychiatry* 197, no. 1 (2010).
14. J. LeDoux, *The Emotional Brain: The Mysterious Underpinnings of Emotional Life* (New York: Simon & Schuster, 1996), 265.
15. R. F. Baumeister and T. F. Heatherton, "Self-Regulation Failure: An Overview," *Psychological Inquiry* 7, no. 1 (1996).

8. PERSONAL VULNERABILITIES AND SUICIDE

1. John Bowlby, *A Secure Base: Clinical Applications of Attachment Theory* (London: Taylor & Francis, 2005).
2. J. G. Allen, P. Fonagy, and A. Bateman, *Mentalizing in Clinical Practice* (Washington, DC: American Psychiatric Publishing, 2008), 20.

3. L. B. Alexander and L. Luborsky, "The Penn Helping Alliance Scales," in *The Psychotherapeutic Process: A Research Handbook*, ed. L. S. Greenberg and W. M. Pinsoff (New York: Guilford, 1986).
4. K. Michel et al., "Therapist Sensitivity Towards Emotional Life-Career Issues and the Working Alliance with Suicide Attempters," *Archives of Suicide Research* 8, no. 3 (2004), https://doi.org/10.1080/13811110490436792.
5. J. Martin, "Cognitive-Meditational Research on Counselling and Psychotherapy," in *Psychotherapy Process Research: Pragmatic and Narrative Approaches*, ed. S. G. Toukmanian and D. L. Rennie (London: Sage, 1992).
6. A. R. Buss, "Causes and Reasons in Attribution Theory: A Conceptual Critique," *Journal of Personality and Social Psychology* 36, no. 11 (1978): 1311–21.
7. K. Michel and L. Valach, "The Narrative Interview with the Suicidal Patient," in *Building a Therapeutic Alliance with the Suicidal Patient*, ed. K. Michel and D. A. Jobes (Washington, DC: APA, 2011).
8. T. Sharot, *The Influential Mind: What the Brain Reveals About Our Power to Change Others* (London: Little, Brown, 2017), 7.
9. R. Charon, *Narrative Medicine: Honoring the Stories of Illness* (New York: Oxford University Press 2006, 2006), vii.

9. A THINK TANK OF CONCERNED THERAPISTS

1. D. A. Jobes, "Collaborating to Prevent Suicide: A Clinical-Research Perspective," *Suicide and Life-Threatening Behavior* 30, no. 1 (2000).
2. J. T. Maltsberger, "2. Empathy and the Historical Context, or How We Learned to Listen to Patients," in *Building a Therapeutic Alliance with the Suicidal Patient*, ed. K. Michel and D. A. Jobes (Washington, DC: APA, 2011).
3. K. Michel et al., "Discovering the Truth in Attempted Suicide," *American Journal of Psychotherapy* 56, no. 3 (2002), http://www.ncbi.nlm.nih.gov/pubmed/12400207.
4. B. Fogarty et al., "Clinicians' Experience of Collaboration in the Treatment of Suicidal Clients Within the Collaborative Assessment and

Management of Suicidality Framework," *Omega* (Westport), May 30, 2021, https://doi.org/10.1177/00302228211020579.
5. J. Hawgood and D. De Leo, "Suicide Prediction—a Shift in Paradigm Is Needed," *Crisis* 37, no. 4 (July 2016), https://doi.org/10.1027/0227-5910/a000440.
6. D. A. Jobes, *Managing Suicidal Risk: A Collaborative Approach*, 2nd ed. (New York: Guilford, 2016).
7. I. Orbach, "Therapeutic Empathy with the Suicidal Wish: Principles of Therapy with Suicidal Individuals," *American Journal of Psychotherapy* 55 (2001).
8. M. D. Rudd, T. Joiner, and M. H. Rajab, *Treating Suicidal Behavior: An Effective, Time-Limited Approach* (New York: Guilford, 2001).
9. G. K. Brown et al., "Cognitive Therapy for the Prevention of Suicide Attempts: A Randomized Controlled Trial," *Journal of the American Medical Association* 294, no. 5 (2005).
10. J. Bowlby, *A Secure Base: Clinical Applications of Attachment Theory* (London: Routledge, 1988).
11. Jeremy Holmes, *The Search for the Secure Base: Attachment Theory and Psychotherapy* (Hove: Brunner-Routledge, 2001).
12. Holmes, *The Search for the Secure Base*.
13. J. G. Allen, P. Fonagy, and A. Bateman, *Mentalizing in Clinical Practice* (Washington, DC: American Psychiatric Publishing, 2008).
14. S. Rizvi, "The Therapeutic Relationship in Dialectical Behaviour Therapy," in *Building a Therapeutic Alliance with the Suicidal Patient*, ed. K. Michel and D. A. Jobes (Washington, DC: APA, 2011).
15. J. R. Rogers and K. M. Soyka, "'One Size Fits All': An Existential-Constructivist Perspective on the Crisis Intervention Approach with Suicidal Individuals," *Journal of Contemporary Psychotherapy* 34, no. 1 (2004).
16. R.A. Young and L. Valach, "The Self-Confrontation Interview in Suicide Research," *Lifenotes* 7 (2002).
17. L. Valach, R. A. Young, and M. J. Lynam, *Action Theory: A Primer for Applied Research in the Social Sciences* (Westport, CT: Praeger, 2002); L. Valach, "Suicide Attempt: Action Theory Informed Case Study," *Psychology and Behavioral Science International Journal* 16, no. 5 (2021), https://doi.org/10.19080/PBSIJ.2021.16.555948; and L. Valach

and R. A. Young, "No Feeling During Repeated Suicide Attempt: A Qualitative Study," *Journal of Psychiatry* 21, no. 5 (2018), https://doi.org/10.4172/2378-5756.1000460.
18. K. Michel and D. A. Jobes, eds., *Building a Therapeutic Alliance with the Suicidal Patient* (Washington, DC: APA, 2011).

10. TRANSLATING ACQUIRED KNOWLEDGE INTO A NEW THERAPY

1. K. Michel and L. Valach, "Suicide as Goal-Directed Action," *Archives of Suicide Research* 3, no. 3 (1997).
2. U. Hepp et al., "Psychological and Psychosocial Interventions After Attempted Suicide: An Overview of Treatment Studies," *Crisis* 25, no. 3 (2004), https://doi.org/10.1027/0227-5910.25.3.108.
3. J. A. Motto and A. G. Bostrom, "A Randomized Controlled Trial of Postcrisis Suicide Prevention," *Psychiatric Services* 52, no. 6 (2001).
4. D. Kahneman, *Thinking Fast and Slow* (New York: Farrar, Strauss and Giroux, 2011), 194.
5. "Kevin Hines, Survivor, Storyteller, Film Maker," https://www.kevinhinesstory.com/.
6. D. Ventrice et al., "Suicide Attempters' Memory Traces of Exposure to Suicidal Behavior: A Qualitative Pilot Study," *Crisis* 31, no. 2 (2010), https://doi.org/10.1027/0227-5910/a000013.
7. Kahneman, *Thinking Fast and Slow*, 417.
8. Y. Dudai, "The Neurobiology of Consolidations, or, How Stable Is the Engram?," *Annual Review of Psychology* 55 (2004), https://doi.org/10.1146/annurev.psych.55.090902.142050.
9. A. Wenzel, G. K. Brown, and A. T. Beck, *Cognitive Therapy for Suicidal Patients: Scientific and Clinical Applications* (Washington, DC: APA, 2009).
10. M. Olfson, "Suicide Risk After Psychiatric Hospital Discharge," *JAMA Psychiatry* 74, no. 7 (July 1, 2017), https://doi.org/10.1001/jamapsychiatry.2017.1043.
11. E. R. Kandel, "A New Intellectual Framework for Psychiatry," *American Journal of Psychiatry* 155, no. 4 (April 1998), https://doi.org/10.1176/ajp.155.4.457.

12. A. Gysin-Maillart and K. Michel, *Kurztherapie nach Suizidversuch ASSIP, Attempted Suicide Short Intervention Program* (Bern: Hans Huber, 2013).
13. K. Michel and A. Gysin-Maillart, *ASSIP—Attempted Suicide Short Intervention Program: A Manual for Clinicians* (Göttingen: Hogrefe, 2015), http://doi.org/10.1027/00476-000.
14. A. Gysin-Maillart et al., "A Novel Brief Therapy for Patients Who Attempt Suicide: A 24-Months Follow-Up Randomized Controlled Study of the Attempted Suicide Short Intervention Program (ASSIP)," *PLoS Med* 13, no. 3 (March 2016), https://doi.org/10.1371/journal.pmed.1001968.
15. F. Naudet et al., "Data Sharing and Reanalysis of Randomized Controlled Trials in Leading Biomedical Journals with a Full Data Sharing Policy: Survey of Studies Published in the *BMJ* and *PLOS Medicine*," *BMJ* 360 (February 13, 2018), https://doi.org/10.1136/bmj.k400.
16. A. L. Park et al., "Cost-Effectiveness of a Brief Structured Intervention Program Aimed at Preventing Repeat Suicide Attempts Among Those Who Previously Attempted Suicide: A Secondary Analysis of the ASSIP Randomized Clinical Trial," *JAMA Network Open* 1, no. 6 (October 5, 2018), https://doi.org/10.1001/jamanetworkopen.2018.3680.
17. R. H. Thaler and C. R. Sunstein, *Nudge: Improving Decisions About Health, Wealth, and Happiness* (New Haven, CT: Yale University Press, 2008).

11. NOW WHAT DOES THIS ALL MEAN FOR SUICIDE PREVENTION?

1. Y. T. Wang et al., "Suicide Misconceptions and Attitudes Toward Suicide Prevention Measures in Taiwan," *Crisis*, November 29, 2022, https://doi.org/10.1027/0227-5910/a000893, https://www.ncbi.nlm.nih.gov/pubmed/36444884.
2. L. P. Astraud, J. A. Bridge, and F. Jollant, "Thirty Years of Publications in Suicidology: A Bibliometric Analysis," *Archives of Suicide Research* 25, no. 4 (October–December 2021), https://doi.org/10.1080/13811118.2020.1746944.

3. H. Brody, "'The Deepest Silences': What Lies Behind the Arctic's Indigenous Suicide Crisis," *Guardian*, July 21, 2022, https://www.theguardian.com/news/2022/jul/21/the-deepest-silences-what-lies-behind-the-indigenous-suicide-crisis.
4. U.S. Department of Health and Human Services, Office of Disease Prevention and Health Promotion, "National Action Plan to Improve Health Literacy," 2010, https://health.gov/our-work/national-health-initiatives/health-literacy/national-action-plan-improve-health-literacy.
5. T. Niederkrotenthaler et al., "Increasing Help-Seeking and Referrals for Individuals at Risk for Suicide by Decreasing Stigma: The Role of Mass Media," *American Journal of Preventitive Medicine* 47, no. 3, suppl 2 (September 2014), https://doi.org/10.1016/j.amepre.2014.06.010.
6. S. J. Fitzpatrick, "Epistemic Justice and the Struggle for Critical Suicide Literacy," *Social Epistemology* 34, no. 6 (2020), https://doi.org/10.1080/02691728.2020.1725921.
7. W. I. Chan et al., "Suicide Literacy, Suicide Stigma and Help-Seeking Intentions in Australian Medical Students," *Australasian Psychiatry* 22, no. 2 (April 2014), https://doi.org/10.1177/1039856214522528.
8. M. Ftanou et al., "Public Service Announcements to Change Attitudes About Youth Suicide: A Randomized Controlled Trial," *Archives of Suicide Research* 25, no. 4 (October–December 2021), https://doi.org/10.1080/13811118.2020.1765929.
9. A. L. Calear et al., "The Literacy of Suicide Scale," *Crisis*, June 15, 2021, https://doi.org/10.1027/0227-5910/a000798.
10. P. J. Batterham et al., "Suicide Stigma and Suicide Literacy in a Clinical Sample," *Suicide and Life Threatening Behavior* 49, no. 4 (August 2019), https://doi.org/10.1111/sltb.12496.
11. R. H. Thaler and C. R. Sunstein, *Nudge: Improving Decisions About Health, Wealth, and Happiness* (New Haven, CT: Yale University Press, 2008).
12. J. Cherkis, "The Best Way to Save People from Suicide," *Huffington Post*, 2018, https://highline.huffingtonpost.com/articles/en/how-to-help-someone-who-is-suicidal/; "Suicide, Veterans, and How a Simple Idea Is Trying to Combat a Crisis," *RETRO REPORT*, https://www.youtube.com/watch?v=8l1HppUUtaQ.

13. K. Michel and L. Valach, "The Narrative Interview with the Suicidal Patient," in *Building a Therapeutic Alliance with the Suicidal Patient*, ed. K. Michel and D. A. Jobes (Washington, DC: APA, 2011).
14. "LivingWorks ASIST," https://www.livingworks.net/asist.
15. "LivingWorks safeTALK," https://www.livingworks.net/safetalk.

12. A SPECIAL CONCERN: YOUNG PEOPLE

1. K. Michel, "Suicide in Young People Is Different," *Crisis* 9, no. 2 (November 1988), https://www.ncbi.nlm.nih.gov/pubmed/3215036.
2. A. W. Tornblom, A. Werbart, and P. A. Rydelius, "Shame Behind the Masks: The Parents' Perspective on Their Sons' Suicide," *Archives of Suicide Research* 17, no. 3 (2013), https://doi.org/10.1080/13811118.2013.805644.
3. Tornblom, Werbart, and Rydelius, "Shame Behind the Masks."
4. S. Curtis et al., "Caring for Young People Who Self-Harm: A Review of Perspectives from Families and Young People," *International Journal of Environmental Research and Public Health* 15, no. 5 (May 10, 2018), https://doi.org/10.3390/ijerph15050950.
5. D. Lester, "Denial in Suicide Survivors," *Crisis* 25, no. 2 (2004), https://doi.org/10.1027/0227-5910.25.2.78.
6. K.A. King et al., "How Confident do High School Counselors Feel in Recognizing Students at Risk for Suicide?," *American Journal of Health and Behavior* 23 (1999).
7. J. Pronovost, "Prevalence of Suicidal Ideations and Behaviors Among Adolescents," in *Suicidal Behavior and Risk Factors*, ed. G. Ferrari, M. Bellini, and P. Crepet (Bologna: Monduzzi Editore, 1990).
8. K. Michel, letter, *Crisis* 26, no. 1 (2005), https://doi.org/10.1027/0227-5910.26.1.36.
9. D. Gillies et al., "Prevalence and Characteristics of Self-Harm in Adolescents: Meta-Analyses of Community-Based Studies 1990–2015," *Journal of the American Academy of Child and Adolescent Psychiatry* 57, no. 10 (October 2018), https://doi.org/10.1016/j.jaac.2018.06.018.
10. R. Wu et al., "A Large Sample Survey of Suicide Risk Among University Students in China," *BMC Psychiatry* 21, no. 1 (September 28, 2021), https://doi.org/10.1186/s12888-021-03480-z.

11. S. Williams, "Denmark Plans Regulation of Influencers Following Suicide Note," BBC, 2019, https://www.bbc.com/news/world-europe-48928012.
12. L. La Sala et al., "Can a Social Media Intervention Improve Online Communication About Suicide? A Feasibility Study Examining the Acceptability and Potential Impact of the #chatsafe Campaign," *PLoS One* 16, no. 6 (2021), https://doi.org/10.1371/journal.pone.0253278.
13. D. Nunez et al., "Testing the Effectiveness of a Blended Intervention to Reduce Suicidal Ideation Among School Adolescents in Chile: A Protocol for a Cluster Randomized Controlled Trial," *International Journal of Environmental Research and Public Health* 19, no. 7 (March 26, 2022), https://doi.org/10.3390/ijerph19073947.
14. K. A. Clark, J. F. Sexton, and T. McKay, "Association of Sexual Orientation with Exposure to Suicide and Related Emotional Distress Among US Adults," *Archives of Suicide Research* (September 27, 2022), https://doi.org/10.1080/13811118.2022.2127386.
15. T. Surace et al., "Lifetime Prevalence of Suicidal Ideation and Suicidal Behaviors in Gender Non-conforming Youths: A Meta-analysis," *European Child & Adolescent Psychiatry* 30, no. 8 (August 2021), https://doi.org/10.1007/s00787-020-01508-5.
16. S. S. Canetto et al., "Suicidal as Normal—a Lesbian, Gay, and Bisexual Youth Script?," *Crisis* 42, no. 4 (Jul 2021), https://doi.org/10.1027/0227-5910/a000730.
17. E. Yard et al., "Emergency Department Visits for Suspected Suicide Attempts Among Persons Aged 12–25 Years Before and During the COVID-19 Pandemic—United States, January 2019–May 2021," *Morbidity and Mortality Weekly Report* 70, no. 24 (June 18, 2021), https://doi.org/10.15585/mmwr.mm7024e1.
18. A. Slomski, "Pediatric Depression and Anxiety Doubled During the Pandemic," *Journal of the American Medical Association* 326, no. 13 (October 5, 2021), https://doi.org/10.1001/jama.2021.16374.
19. E. Balt et al., "Gender Differences in Suicide-Related Communication of Young Suicide Victims," *PLoS One* 16, no. 5 (2021), https://doi.org/10.1371/journal.pone.0252028.
20. M. L. Rasmussen, H. Haavind, and G. Dieserud, "Young Men, Masculinities, and Suicide," *Archives of Suicide Research* 22, no. 2 (April–June 2018), https://doi.org/10.1080/13811118.2017.1340855.

12. A SPECIAL CONCERN: YOUNG PEOPLE ☙ 317

21. A. Bradley-Ewing et al., "Parent and Adolescent Thoughts About Suicide Risk Screening in Pediatric Outpatient Settings," *Archives of Suicide Research* 26, no. 3 (July–September 2022), https://doi.org/10.1080/13811118.2020.1864536.
22. I. Bellairs-Walsh et al., "Best Practice When Working with Suicidal Behaviour and Self-Harm in Primary Care: A Qualitative Exploration of Young People's Perspectives," *BMJ Open* 10, no. 10 (October 28, 2020), https://doi.org/10.1136/bmjopen-2020-038855.
23. D. Wasserman et al., "School-Based Suicide Prevention Programmes: The SEYLE Cluster-Randomised, Controlled Trial," *Lancet* 385, no. 9977 (April 18, 2015), https://doi.org/10.1016/S0140-6736(14)61213-7.
24. N. Buus et al., "Stakeholder Perspectives on Using and Developing the MYPLAN Suicide Prevention Mobile Phone Application: A Focus Group Study," *Archives of Suicide Research* 24, no. 1 (January–March 2020), https://doi.org/10.1080/13811118.2018.1489319.
25. Q. Cheng et al., "Co-Creation and Impacts of a Suicide Prevention Video," *Crisis* 41, no. 1 (January 2020), https://doi.org/10.1027/0227-5910/a000593; "Suicide Prevention Short Film," https://www.youtube.com/watch?v=8tnfiKX1sc8&t=797s.
26. E. Berger, P. Hasking, and G. Martin, "'Listen to Them': Adolescents' Views on Helping Young People Who Self-Injure," *Journal of Adolescence* 36, no. 5 (October 2013), https://doi.org/10.1016/j.adolescence.2013.07.011.
27. Curtis et al., "Caring for Young People Who Self-Harm."
28. E. Duarte et al., "How Do Families Represent the Functions of Deliberate Self-Harm? A Comparison Between the Social Representations from Adolescents and Their Parents," *Archives of Suicide Research* 24, no. sup1 (2020), https://doi.org/10.1080/13811118.2018.1545713.
29. C. A. King et al., "The Youth-Nominated Support Team-Version II for Suicidal Adolescents: A Randomized Controlled Intervention Trial," *Journals of Consulting and Clinical Psychology* 77, no. 5 (October 2009), https://doi.org/10.1037/a0016552.
30. D. A. Ruch et al., "Trends in Suicide Among Youth Aged 10 to 19 Years in the United States, 1975 to 2016," *JAMA Netw Open* 2, no. 5 (May 3, 2019), https://doi.org/10.1001/jamanetworkopen.2019.3886.
31. J. M. Twenge, *iGen* (New York: ATRIA, 2017).

32. J. Luby and S. Kertz, "Increasing Suicide Rates in Early Adolescent Girls in the United States and the Equalization of Sex Disparity in Suicide: The Need to Investigate the Role of Social Media," *JAMA Netw Open* 2, no. 5 (May 3 2019), https://doi.org/10.1001/jamanetworkopen.2019.3916.
33. M. M. Husky et al., "Bullying Involvement and Suicidal Ideation in Elementary School Children Across Europe," *Journal of Affective Disorders* 299 (February 15, 2022), https://doi.org/10.1016/j.jad.2021.12.023.
34. M. Dean, "The Story of Amanda Todd," *New Yorker*, October 18, 2012, https://www.newyorker.com/culture/culture-desk/the-story-of-amanda-todd.
35. J. N. Sales, *American Girls: Social Media and the Secret Lives of Teenagers* (New York: Knopf Doubleday, 2016), 106.
36. A. Satariano and M. Isaac, "Facebook Whistle-Blower Brings Campaign to Europe After Disclosures," *New York Times*, October 25 2021, https://www.nytimes.com/2021/10/25/business/frances-haugen-facebook.html.
37. J. A. Bridge et al., "Association Between the Release of Netflix's *13 Reasons Why* and Suicide Rates in the United States: An Interrupted Time Series Analysis," *Journal of the American Academy of Child and Adolescent Psychiatry* 59, no. 2 (February 2020), https://doi.org/10.1016/j.jaac.2019.04.020.
38. R. Voelker, "Mounting Evidence and Netflix's Decision to Pull a Controversial Suicide Scene," *Journal of the American Medical Association* 322, no. 6 (August 13, 2019), https://doi.org/10.1001/jama.2019.9492.
39. Bridge et al., "Association Between the Release of Netflix's *13 Reasons Why* and Suicide Rates in the United States."
40. D. Reidenberg et al., "*13 Reasons Why*: The Evidence Is in and Cannot Be Ignored," *Journal of the American Academy of Child and Adolescent Psychiatry* 59, no. 9 (September 2020), https://doi.org/10.1016/j.jaac.2020.01.019.
41. A. Schmidtke and H. Häfner, "Die Vermittlung von Selbstmordmotivation und Selbstmordhandlung durch fiktive Modelle. Die Folgen der Fernsehserie 'Tod eines Schülers,'" *Nervenarzt* 57 (1986).

42. A. Bandura, *Social Learning Theory* (Englewood Cliffs, NJ: Prentice Hall, 1977).
43. D. P. Phillips, K. Lesyna, and D. J. Paight, "Suicide and the Media," in *Assessment and Prediction of Suicide*, ed. R. W. Maris, A. L. Berman, and J. T. Maltsberger (New York: Guilford, 1992).
44. J. Berger, *Contagious: Why Things Catch On* (New York: Simon & Schuster, 2013), 152.

13. FOR HEALTH PROFESSIONALS: IT'S ABOUT THE PERSON IN THE PATIENT

1. E. F. Ward-Ciesielski and S. L. Rizvi, "The Potential Iatrogenic Effects of Psychiatric Hospitalization for Suicidal Behavior: A Critical Review and Recommendations for Research," *Clinical Psychology Science and Practice*, 2020, https://doi.org/https://doi.org/10.1111/cpsp.12332.
2. P. Quinnett, "The Role of Clinician Fear in Interviewing Suicidal Patients," *Crisis* 40, no. 5 (September 2019), https://doi.org/10.1027/0227-5910/a000555.
3. T. Podlogar et al., "The Model of Dynamic Balance in Therapists' Experiences and Views on Working with Suicidal Clients: A Qualitative Study," *Clinical Psychology and Psychotherapy* 27, no. 6 (November 2020), https://doi.org/10.1002/cpp.2484.
4. K. Michel et al., "Discovering the Truth in Attempted Suicide," *American Journal of Psychotherapy* 56, no. 3 (2002), http://www.ncbi.nlm.nih.gov/pubmed/12400207.
5. T. E. Ellis et al., "Impact of a Suicide-Specific Intervention Within Inpatient Psychiatric Care: The Collaborative Assessment and Management of Suicidality," *Suicide and Life Threatening Behavior* 45, no. 5 (October 2015), https://doi.org/10.1111/sltb.12151.
6. J. Hawgood et al., "Gatekeeper Training and Minimum Standards of Competency," *Crisis*, June 30, 2021, https://doi.org/10.1027/0227-5910/a000794.
7. I. Bellairs-Walsh et al., "Working with Young People at Risk of Suicidal Behaviour and Self-Harm: A Qualitative Study of Australian General Practitioners' Perspectives," *International Journal of*

Environmental Research and Public Health 18, no. 24 (December 8, 2021), https://doi.org/10.3390/ijerph182412926.

8. Selma Gaily-Luoma et al., "How Do Health Care Services Help and Hinder Recovery After a Suicide Attempt? A Qualitative Analysis of Finnish Service User Perspectives," *International Journal of Mental Health Systems* 16, no. 1 (2022), https://doi.org/10.1186/s13033-022-00563-6.

9. D. Murray and P. Devitt, "Suicide Risk Assessment Doesn't Work," *Scientific American*, March 28, 2017, https://www.scientificamerican.com/article/suicide-risk-assessment-doesnt-work/.

10. M. K. Chan et al., "Predicting Suicide Following Self-Harm: Systematic Review of Risk Factors and Risk Scales," *British Journal of Psychiatry* 209, no. 4 (October 2016), https://doi.org/10.1192/bjp.bp.115.170050.

11. A. L. Berman, "Book Review: The Neuroscience of Suicidal Behavior," *Crisis* 40, no. 4 (2019), https://doi.org/10.1027/0227-5910/a000585.

12. B. E. Belsher et al., "Prediction Models for Suicide Attempts and Deaths: A Systematic Review and Simulation," *JAMA Psychiatry* 76, no. 6 (June 1, 2019), https://doi.org/10.1001/jamapsychiatry.2019.0174.

13. E. D. Klonsky, T. Dixon-Luinenburg, and A. M. May, "The Critical Distinction Between Suicidal Ideation and Suicide Attempts," *World Psychiatry* 20, no. 3 (October 2021), https://doi.org/10.1002/wps.20909.

14. C. R. Glenn and M. K. Nock, "Improving the Short-Term Prediction of Suicidal Behavior," *American Journal of Preventitive Medicine* 47, no. 3 suppl 2 (September 2014), https://doi.org/10.1016/j.amepre.2014.06.004.

15. M. D. Rudd, "Fluid Vulnerability Theory: A Cognitive Approach to Understanding the Process of Acute and Chronic Risk," in *Cognition and Suicide: Theory, Research, and Therapy*, ed. T. E. Ellis (Washington, DC: American Psychological Association, 2006).

16. C. J. Bryan and M. D. Rudd, "The Importance of Temporal Dynamics in the Transition from Suicidal Thought to Behavior," *Clinical Psychology: Science and Practice* 23, no. 1 (2016).

17. G. R. Jenkins et al., "Suicide Rate 22 Years After Parasuicide: Cohort Study," Research Support, Non-U.S. Gov't, *BMJ* 325, no. 7373

(November 16, 2002), http://www.ncbi.nlm.nih.gov/pubmed/12433767; A. L. Beautrais, "Further Suicidal Behavior Among Medically Serious Suicide Attempters," *Suicide and Life-Threatening Behavior* 34, no. 1 (Spring 2004), http://www.ncbi.nlm.nih.gov/pubmed/15106883.

18. M. B. Warren and L. A. Smithkors, "Suicide Prevention in the U.S. Department of Veterans Affairs: Using the Evidence Without Losing the Narrative," *Psychiatric Services* 71, no. 4 (April 1, 2020), https://doi.org/10.1176/appi.ps.201900482.

19. R. McCabe et al., "How Do Healthcare Professionals Interview Patients to Assess Suicide Risk?," *BMC Psychiatry* 17, no. 1 (April 4, 2017), https://doi.org/10.1186/s12888-017-1212-7.

20. Quinnett, "The Role of Clinician Fear in Interviewing Suicidal Patients."

21. K. A. Busch, J. Fawcett, and D. G. Jacobs, "Clinical Correlates of Inpatient Suicide," *Journal of Clinical Psychiatry* 64, no. 1 (2003).

22. M. David Rudd, Michael Mandrusiak, and Thomas E. Joiner Jr., "The Case Against No-Suicide Contracts: The Commitment to Treatment Statement as a Practice Alternative," *Journal of Clinical Psychology* 62, no. 2 (2006).

23. J. R. Cutcliffe et al., "Understanding the Risks of Recent Discharge: The Phenomenological Lived Experiences," *Crisis* 33, no. 1 (January 1, 2012), https://doi.org/10.1027/0227-5910/a000096.

24. Cutcliffe et al., "Understanding the Risks of Recent Discharge."

25. T. Teismann, T. Forkmann, and H. Glaesmer, "Risikoabschätzung bei suizidalen Patienten: Geht das überhaupt?," *Verhaltenstherapie* 29, no. 2 (June 2019), https://doi.org/10.1159/000493887; translation mine.

26. K. Michel, "Suicide Risk Assessment Must Be Collaborative," *Acta Scientific Medical Sciences* 4, no. 12 (2020), https://doi.org/10.31080/ASMS.2020.04.0793.

27. D. A. Jobes, *Managing Suicidal Risk: A Collaborative Approach*, 2nd ed. (New York: Guilford, 2016).

28. J. Peterson, J. Skeem, and S. Manchak, "If You Want to Know, Consider Asking: How Likely Is It That Patients Will Hurt Themselves in the Future?," *Psychological Assessment* 23, no. 3 (2011).

29. E. K. Czyz, A. G. Horwitz, and C. A. King, "Self-Rated Expectations of Suicidal Behavior Predict Future Suicide Attempts Among

Adolescent and Young Adult Psychiatric Emergency Patients," *Depression and Anxiety* 33, no. 6 (June 2016), https://doi.org/10.1002/da.22514.

30. M. Slade, "Implementing Shared Decision Making in Routine Mental Health Care," *World Psychiatry* 16, no. 2 (June 2017), https://doi.org/10.1002/wps.20412.

31. S. Barzilay et al., "Determinants and Predictive Value of Clinician Assessment of Short-Term Suicide Risk," *Suicide and Life Threatening Behavior* 49, no. 2 (April 2019), https://doi.org/10.1111/sltb.12462.

32. M. Hawes et al., "Anhedonia and Suicidal Thoughts and Behaviors in Psychiatric Outpatients: The Role of Acuity," *Depression and Anxiety* 35, no. 12 (December 2018), https://doi.org/10.1002/da.22814.

33. S. Barzilay et al., "Assessment of Near-Term Risk for Suicide Attempts Using the Suicide Crisis Inventory," *Journal of Affective Disorders* 276 (November 1, 2020), https://doi.org/10.1016/j.jad.2020.06.053.

34. J. T. Maltsberger and D. H. Buie, "Countertransference Hate in the Treatment of Suicidal Patients," in *Essential Papers on Suicide*, ed. John T. Maltsberger and Mark J. Goldblatt (New York: New York University Press, 1996).

35. M. F. Myers, *Becoming a Doctors' Doctor: A Memoir* (n.p., 2020), https://www.michaelfmyers.com/becoming_a_doctors__doctor___br__a_memoir_.htm.

36. R. F. Baumeister and T. F. Heatherton, "Self-Regulation Failure: An Overview," *Psychological Inquiry* 7, no. 1 (1996).

37. E. L. Meerwijk et al., "Direct Versus Indirect Psychosocial and Behavioural Interventions to Prevent Suicide and Suicide Attempts: A Systematic Review and Meta-analysis," *Lancet Psychiatry* 3, no. 6 (June 2016), https://doi.org/10.1016/S2215-0366(16)00064-X; T. Sobanski et al., "Psychotherapeutic Interventions for the Prevention of Suicide Re-attempts: A Systematic Review," *Psychological Medicine* (October 5, 2021), https://doi.org/10.1017/S0033291721003081.

38. P. Omerov and J. Bullington, "Nursing Care of the Suicidal Patient," in *Suicide Risk Assessment and Prevention*, ed. M. Pompili (Cham: Springer, 2021).

39. S. P. Thomas, M. Shattell, and T. Martin, "What's Therapeutic About the Therapeutic Milieu?," *Archives of Psychiatric Nursing* 16, no. 3 (2002), https://doi.org/10.1053/apnu.2002.32945.

40. B. A. Arnow and D. Steidtmann, "Harnessing the Potential of the Therapeutic Alliance," *World Psychiatry* 13, no. 3 (2014).
41. I. W. Miller et al., "Suicide Prevention in an Emergency Department Population: The ED-SAFE Study," *JAMA Psychiatry* 74, no. 6 (June 1, 2017), https://doi.org/10.1001/jamapsychiatry.2017.0678; I. W. Miller, B. A. Gaudiano, and L. M. Weinstock, "The Coping Long Term with Active Suicide Program: Description and Pilot Data," *Suicide and Life Threatening Behavior* 46, no. 6 (December 2016), https://doi.org/10.1111/sltb.12247.
42. J. Sinclair and J. Green, "Understanding Resolution of Deliberate Self Harm: Qualitative Interview Study of Patients' Experiences," *BMJ* 330, no. 7500 (2005).
43. L. Luborsky et al., "The Revised Helping Alliance Questionnaire (HAQ-II): Psychometric Properties," *Journal of Psychotherapy Practice and Research* 5 (1996); R. L. Hatcher and J. A. Gillaspy, "Development and Validation of a Revised Short Version of the Working Alliance Inventory," *Psychotherapy Research* 16, no. 1 (2006), https://doi.org/10.1080/10503300500352500.

BIBLIOGRAPHY

Allen, J. G., P. Fonagy, and A. Bateman. *Mentalizing in Clinical Practice.* Washington, DC: American Psychiatric Publishing, 2008.

American Psychiatric Association. *The Diagnostic and Statistical Manual of Mental Disorders.* 5th ed. Washington, DC: American Psychiatric Publishing, 2013.

Angus, L. E., and J. Mcleod. *The Handbook of Narrative and Psychotherapy: Practice, Theory, and Research.* Thousand Oaks, CA: Sage, 2004.

Balint, M. "The Doctor, His Patient, and the Illness." *Lancet* 268, no. 6866 (April 1955): 683–68. https://doi.org/10.1016/s0140-6736(55)91061-8.

Bandura, A. *Social Learning Theory.* Englewood Cliffs, NJ: Prentice Hall, 1977.

Baumeister, R. F. "Suicide as Escape from Self." *Psychological Review* 97, no. 1 (1990): 90–113.

Baumeister, R. F., and T. F. Heatherton. "Self-Regulation Failure: An Overview." *Psychological Inquiry* 7, no. 1 (1996): 1–15.

Bellairs-Walsh, I., Y. Perry, K. Krysinska, et al. "Best Practice When Working with Suicidal Behaviour and Self-Harm in Primary Care: A Qualitative Exploration of Young People's Perspectives." *BMJ Open* 10, no. 10 (October 2020): e038855. https://doi.org/10.1136/bmjopen-2020-038855.

Bolger, E. "Grounded Theory Analysis of Emotional Pain." *Psychotherapy Research* 9, no. 3 (1999): 342–62. https://doi.org/10.1080/10503309912331332801.

Bowlby, J. *A Secure Base. Clinical Applications of Attachment Theory.* London: Routledge, 1988.

Bridge, J. A., J. B. Greenhouse, D. Ruch, et al. "Association Between the Release of Netflix's *13 Reasons Why* and Suicide Rates in the United States: An Interrupted Time Series Analysis." *Journal of the American Academy of Child and Adolescent Psychiatry* 59, no. 2 (February 2020): 236–43. https://doi.org/10.1016/j.jaac.2019.04.020.

Bruner, J. *Making Stories: Law, Literature, Life.* Cambridge, MA: Harvard University Press, 2002.

———. "The Narrative Creation of Self." In *The Handbook of Narrative and Psychotherapy. Practice, Theory, and Research*, ed. L. E. Angus and J. Mcleod, 3–14. Thousand Oaks, CA: Sage, 2004.

Busch, K. A., J. Fawcett, and D. G. Jacobs. "Clinical Correlates of Inpatient Suicide." *Journal of Clinical Psychiatry* 64, no. 1 (2003): 14–19.

Chan, W. I., P. Batterham, H. Christensen, and C. Galletly. "Suicide Literacy, Suicide Stigma and Help-Seeking Intentions in Australian Medical Students." *Australas Psychiatry* 22, no. 2 (April 2014): 132–39. https://doi.org/10.1177/1039856214522528.

Charon, R. *Narrative Medicine: Honoring the Stories of Illness.* New York: Oxford University Press, 2006.

Cherkis, J. "The Best Way to Save People from Suicide." *Huffington Post*, 2018. https://highline.huffingtonpost.com/articles/en/how-to-help-someone-who-is-suicidal/.

De Leo, D. "Suicide Prevention Is Far More Than a Psychiatric Business." *World Psychiatry* 3, no. 3 (October 2004): 155–56. http://www.ncbi.nlm.nih.gov/pubmed/16633482.

Fitzpatrick, S. J. "Epistemic Justice and the Struggle for Critical Suicide Literacy." *Social Epistemology* 34, no. 6 (2020): 555–56. https://doi.org/10.1080/02691728.2020.1725921.

Gysin-Maillart, A., S. Schwab, L. Soravia, et al. "A Novel Brief Therapy for Patients Who Attempt Suicide: A 24-Months Follow-up Randomized Controlled Study of the Attempted Suicide Short Intervention Program (Assip)." *PLoS Med* 13, no. 3 (March 2016): e1001968. https://doi.org/10.1371/journal.pmed.1001968.

Harris, E. C., and B. Barraclough. "Suicide as an Outcome for Mental Disorders. A Meta-Analysis." *British Journal of Psychiatry* 170 (March 1997): 205–28. https://doi.org/10.1192/bjp.170.3.205.

Hawgood, J., and D. De Leo. "Suicide Prediction—a Shift in Paradigm Is Needed." *Crisis* 37, no. 4 (July 2016): 251–55. https://doi.org/10.1027/0227-5910/a000440.

Hepp, U., L. Wittmann, U. Schnyder, and K. Michel. "Psychological and Psychosocial Interventions After Attempted Suicide." *Crisis* 25, no. 3 (2004): 108–17. https://doi.org/10.1027/0227-5910.25.3.108.

Hines, K. "Kevin Hines, Survivor, Storyteller, Film Maker." https://www.kevinhinesstory.com/.

Holmes, J. *The Search for the Secure Base: Attachment Theory and Psychotherapy*. Hove: Brunner Routledge, 2001.

Husky, M. M., A. Bitfoi, M. G. Carta, et al. "Bullying Involvement and Suicidal Ideation in Elementary School Children Across Europe." *Journal of Affective Disorders* 299 (February 15, 2022): 281–86. https://doi.org/10.1016/j.jad.2021.12.023.

Jaspers, K. *General Psychopathology*. Trans. J. Hoenig and M. Hamilton. Manchester: Manchester University Press, 1963.

Jobes, D. A. "Collaborating to Prevent Suicide: A Clinical-Research Perspective." *Suicide and Life-Threatening Behavior* 30, no. 1 (2000): 8–17.

———. *Managing Suicidal Risk: A Collaborative Approach*. 2nd ed. New York: Guilford, 2016.

Joiner Jr., T. E., K. A. Van Orden, T. K. Witte, and M. D. Rudd. *The Interpersonal Theory of Suicide: Guidance for Working with Suicidal Clients*. Washington, DC: American Psychological Association, 2009.

Jollant, F., F. Bellivier, M. Leboyer, et al. "Impaired Decision Making in Suicide Attempters." *American Journal of Psychiatry* 162, no. 2 (2005): 304–10.

Kahneman, D. *Thinking Fast and Slow*. New York: Farrar, Straus and Giroux, 2011.

Kandel, E. R. "A New Intellectual Framework for Psychiatry." *American Journal of Psychiatry* 155, no. 4 (April 1998): 457–69. https://doi.org/10.1176/ajp.155.4.457.

La Sala, L., Z. Teh, M. Lamblin, et al. "Can a Social Media Intervention Improve Online Communication About Suicide? A Feasibility Study Examining the Acceptability and Potential Impact of the #Chatsafe Campaign." *PLoS One* 16, no. 6 (2021): e0253278. https://doi.org/10.1371/journal.pone.0253278.

LeDoux, J. *The Emotional Brain: The Mysterious Underpinnings of Emotional Life*. New York: Simon & Schuster, 1996.

Libet, B., C. A. Gleason, E. W. Wright, and D. K. Pearl. "Time of Conscious Intention to Act in Relation to Onset of Cerebral Activity (Readiness-Potential). The Unconscious Initiation of a Freely Voluntary Act." *Brain* 106 (part 3) (September 1983): 623–42. https://doi.org/10.1093/brain/106.3.623.

Linehan, M. M. *Cognitive Behavioural Therapy of Borderline Personality Disorder*. New York: Guilford, 1993.

Lizardi, D., and B. Stanley. "Treatment Engagement: A Neglected Aspect in the Psychiatric Care of Suicidal Patients." *Psychiatric Services* 61, no. 12 (2010): 1183–91.

Luby, J., and S. Kertz. "Increasing Suicide Rates in Early Adolescent Girls in the United States and the Equalization of Sex Disparity in Suicide: The Need to Investigate the Role of Social Media." *JAMA Netw Open* 2, no. 5 (May 3 2019): e193916. https://doi.org/10.1001/jamanetworkopen.2019.3916.

Maris, Ronald W., and Bernard Melvin Lazerwitz. *Pathways to Suicide: A Survey of Self-Destructive Behaviors*. Baltimore: Johns Hopkins University Press, 1981.

McCabe, R., I. Sterno, S. Priebe, et al. "How Do Healthcare Professionals Interview Patients to Assess Suicide Risk?" *BMC Psychiatry* 17, no. 1 (April 4 2017): 122. https://doi.org/10.1186/s12888-017-1212-7.

Meerwijk, E. L., and S. J. Weiss. "Toward a Unifying Definition of Psychological Pain." *Journal of Loss & Trauma* 16, no. 5 (2011): 402–12. https://doi.org/10.1080/15325024.2011.572044.

Michel, K. "Suicide Risk Assessment Must Be Collaborative." *Acta Scientific Medical Sciences* 4, no. 12 (2020): 38–43. https://doi.org/10.31080/ASMS.2020.04.0793.

Michel, K., P. Dey, K. Stadler, and L. Valach. "Therapist Sensitivity Towards Emotional Life-Career Issues and the Working Alliance with Suicide Attempters." *Archives of Suicide Research* 8, no. 3 (2004): 203–13. https://doi.org/10.1080/13811110490436792.

Michel, K., and A. Gysin-Maillart. *ASSIP—Attempted Suicide Short Intervention Program. A Manual for Clinicians*. Göttingen: Hogrefe, 2015. doi:http://doi.org/10.1027/00476-000.

Michel, K., A. Gysin-Maillart, S. Breit, et al. "Psychopharmacological Treatment Is Not Associated with Reduced Suicide Ideation and Reattempts in an Observational Follow-up Study of Suicide Attempters." *Journal of Psychiatric Research* 140 (August 2021): 180–86. https://doi.org/10.1016/j.jpsychires.2021.05.068.

Michel, K., and D. A. Jobes, eds. *Building a Therapeutic Alliance with the Suicidal Patient.* Washington, DC: APA, 2011.

Michel, K., J. T. Maltsberger, D. A. Jobes, et al. "Discovering the Truth in Attempted Suicide." Case Reports. *American Journal of Psychotherapy* 56, no. 3 (2002): 424–37. http://www.ncbi.nlm.nih.gov/pubmed/12400207.

Michel, K., and L. Valach. "The Narrative Interview with the Suicidal Patient." In *Building a Therapeutic Alliance with the Suicidal Patient*, ed. K. Michel and D. A. Jobes, 63–80. Washington, DC: APA, 2011.

Michel, K., and L. Valach. "Suicide as Goal-Directed Action." *Archives of Suicide Research* 3, no. 3 (1997): 213–21.

Miller, I. W., C. A. Camargo Jr., S. A. Arias, et al. "Suicide Prevention in an Emergency Department Population: The Ed-Safe Study." *JAMA Psychiatry* 74, no. 6 (June 1, 2017): 563–70. https://doi.org/10.1001/jamapsychiatry.2017.0678.

Murray, D., and P. Devitt. "Suicide Risk Assessment Doesn't Work." *Scientific American*, March 28, 2017. https://www.scientificamerican.com/article/suicide-risk-assessment-doesnt-work/.

Myers, M. F. *Becoming a Doctors' Doctor: A Memoir.* 2020. https://www.michaelfmyers.com/becoming_a_doctors__doctor___br__a_memoir_.htm.

NHS Foundation Trust. "Prescribing Guidelines for the Management of Depression." 2021. https://www.candi.nhs.uk/sites/default/files/Depression%20%20Prescribing%20Guidelines%202021.pdf.

Orbach, I. "Dissociation, Physical Pain, and Suicide: A Hypothesis." Review. *Suicide and Life-Threatening Behavior* 24, no. 1 (Spring 1994): 68–79. http://www.ncbi.nlm.nih.gov/pubmed/8203010.

Park, A. L., A. Gysin-Maillart, T. J. Muller, et al. "Cost-Effectiveness of a Brief Structured Intervention Program Aimed at Preventing Repeat Suicide Attempts Among Those Who Previously Attempted Suicide: A Secondary Analysis of the Assip Randomized Clinical Trial." *JAMA Network Open* 1, no. 6 (October 5, 2018): e183680. https://doi.org/10.1001/jamanetworkopen.2018.3680.

Quinnett, P. "The Role of Clinician Fear in Interviewing Suicidal Patients." *Crisis* 40, no. 5 (September 2019): 355–59. https://doi.org/10.1027/0227-5910/a000555.

Rasmussen, M. L., H. Haavind, and G. Dieserud. "Young Men, Masculinities, and Suicide." *Archives of Suicide Research* 22, no. 2 (April–June 2018): 327–43. https://doi.org/10.1080/13811118.2017.1340855.

Reisch, T., E. Seifritz, F. Esposito, et al. "An fMri Study on Mental Pain and Suicidal Behavior." *Journal of Affective Disorders* 126, no. 1 (2010): 321–25.

Rogers, J. R., and K. M. Soyka. "'One Size Fits All': An Existential-Constructivist Perspective on the Crisis Intervention Approach with Suicidal Individuals." *Journal of Contemporary Psychotherapy* 34, no. 1 (2004): 7–22.

Rudd, M. D. "Fluid Vulnerability Theory: A Cognitive Approach to Understanding the Process of Acute and Chronic Risk." In *Cognition and Suicide: Theory, Research, and Therapy*, ed. T. E. Ellis, 355–68. Washington, DC: American Psychological Association, 2006.

Rudd, M. D. "The Suicidal Mode: A Cognitive-Behavioral Model of Suicidality." *Suicide and Life-Threatening Behavior* 30, no. 1 (2000): 18–33.

Rudd, M. D., and G. K. Brown. "A Cognitive Theory of Suicide: Building Hope in Treatment and Strengthening the Therapeutic Relationship." In *Building a Therapeutic Alliance with the Suicidal Patient*, ed. K. Michel and D. A. Jobes, 169–82. Washington, DC: APA, 2011.

Rudd, M. D., T. Joiner, and M. H. Rajab. *Treating Suicidal Behavior: An Effective, Time-Limited Approach*. New York: Guilford, 2001.

Seeger, P., dir. *In Search of Memory: The Neuroscientist Eric Kandel*. 2009. https://icarusfilms.com/if-mem.

Sharot, T. *The Influential Mind: What the Brain Reveals About Our Power to Change Others*. London: Little, Brown, 2017.

Shneidman, E.S. *Definition of Suicide*. New York: Wiley, 1985.

Shneidman, E. S., and N. L. Farberow. *Clues to Suicide*. New York: McGraw-Hill, 1957.

Soper, C. A. *The Evolution of Suicide*. Ed. T. K. Shackelford and V. A. Weekes-Shackelford. Springer, 2018.

Stanley, B., and J. J. Mann. "The Need for Innovation in Health Care Systems to Improve Suicide Prevention." *JAMA Psychiatry* 77, no. 1 (January 1, 2020): 96–98. https://doi.org/10.1001/jamapsychiatry.2019.2769.

Stefan, S. *Rational Suicide, Irrational Laws.* New York: Oxford University Press, 2016.
Thaler, R. H., and C. R. Sunstein. *Nudge: Improving Decisions About Health, Wealth, and Happiness.* New Haven, CT: Yale University Press, 2008.
Tornblom, A. W., A. Werbart, and P. A. Rydelius. "Shame Behind the Masks: The Parents' Perspective on Their Sons' Suicide." *Archives of Suicide Research* 17, no. 3 (2013): 242–61. https://doi.org/10.1080/13811118.2013.805644.
Twenge, J. M. *Igen.* New York: ATRIA, 2017.
Valach, L., R. A. Young, and M. J. Lynam. *Action Theory: A Primer for Applied Research in the Social Sciences.* Westport, CT: Praeger, 2002.
van Os, J., and D. Kamp. "Putting the Psychotherapy Spotlight Back on the Self-Reflecting Actors Who Make It Work." *World Psychiatry* 18, no. 3 (October 2019): 292–93. https://doi.org/10.1002/wps.20667.
Warren, M. B., and L. A. Smithkors. "Suicide Prevention in the U.S. Department of Veterans Affairs: Using the Evidence Without Losing the Narrative." *Psychiatric Services* 71, no. 4 (April 1, 2020): 398–400. https://doi.org/10.1176/appi.ps.201900482.
Wegner, D. M. *The Illusion of Conscious Will.* Cambridge, MA: MIT Press, 2002.
Wenzel, A., G. K. Brown, and A. T. Beck. *Cognitive Therapy for Suicidal Patients: Scientific and Clinical Applications.* Washington, DC: American Psychological Association, 2009.
Williams, M. *Suicide and Attempted Suicide.* 2nd ed. London: Penguin, 2002.

INDEX

abandonment, 67, 129, 142, 172, 235; separation anxiety and, 38–40, 132, 140–43, 149–53, 195
action theory, 111–14, 164, 177
acute strategies, 195
AdoASSIP, 205, 233
adolescents. *See* youth
adrenaline, 69, 133
advice giving, 206, 288
Aeschi Working Group: Aeschi Guidelines for Clinicians, 252–54; "Aeschi philosophy" and, 177–79, 273; self-confrontation and video playback in, 163–68; therapeutic relationship and, 162–79; on understanding suicide attempts, 162–79
AI. *See* artificial intelligence
alarm system, 133–35, 278
alcohol, 15, 277
algorithms, 258–61
alienation, 97
Allen, Jon G., 174–75

American Association of Suicidology (AAS), 161–64, 258
American Journal of Psychiatry, 57–58, 169
amygdala, 43, 43–44, 66–69, 130, 133–35, 178
animal behavior, 95, 134
antidepressant medications: brain and, 71–79; depression and, 71–79, 81–82, 107–8; guidelines for using, 75; increased suicidal behavior and, 73–75; media on, 75; pharmaceutical industry and, 71–74; placebo effect of, 75–76; serotonin and, 72; suicide prevention and, 71–79, 81–82
Aplysia californica (snail), 57–58
arrested flight, 95
artificial intelligence (AI), 258–61
Åsberg, Mary, 62
ASSIP. *See* Attempted Suicide Short Intervention Program
attachment theory, 115–17, 148, 174, 182, 233–34

attempted suicide. *See* suicide attempts
Attempted Suicide Short Intervention Program (ASSIP), xv, 120, 270; AdoASSIP, 205, 233; collaborative psychoeducation and, 190–91; components of, 184–200; helpful long-term measures, 194; homework task and, 184, 191; joint case conceptualization and, 191–94; letters and, 197–200, 203; mini-exposure sessions in, 197; narrative interviews in, 185–86; patient handout, 275–79; randomized controlled study and publishing of, 200–205; safety planning in, 194–97, 196; on suicide risk factors, 275–79; treatment model and diagram, 195, 195–96; video playback in, 186–90; warning signs in, 194–95
attentive silence, 148
attitude, toward help, 288–89
autonomy, 206, 218, 288–89
autopilot mode, xiii, 47–53. *See also* dissociation
awareness: self-awareness about suicidal thoughts, 212–13; for suicide prevention, 241–43

Balint, Michael, 89
Bancroft et al., 30–33
Baumeister, Roy F., 94–95, 125, 132, 135, 267

Beck, Aaron, 49–50, 96
behavior: neurobiology and, 57–58, 130–31; reason-cause discussion on, 156. *See also* suicidal behavior
bereaved parents, 126–28; guilt and, 230; personal message to, 230–32; understanding reasons for suicide and, 128, 224–26
Berger, Gregor, 234
Berman, Lanny, 163–64, 249, 258
Bertolote et al., 75
Beyond Blue, 240
Bille-Brahe, Unni, 28
biographical context: case vignettes on, 107–10; understanding and communication in, 107–21
bipolar disorder, 77
Bleuler, Eugen, 13
Blick (magazine), 24–26, 25
bodily functions, 50
Bolger, E., 129
borderline personality disorder (BPD), 97, 174–75
Bostwick, Michael, 169, 172
Bowlby, John, 115–16, 117, 174
BPD. *See* borderline personality disorder
brain: amygdala, 43, 43–44, 68–69, 133–35; antidepressant medications and, 71–79; behavior and, 57–58, 130–31; Brodmann Area 46, 45–46, 46, 66; cerebrospinal fluid and, 62–63; cognitive reframing and,

68; cortex and, 65–66; decision making and, 65; depression and, 69–79; dual processing theory and, 186, 187–89, 194–95; EDA and, 37–40, 39, 44; emotional, xiv, 130–31, 133–35; emotions and, 58, 62–68, 211–13; fMRI and, 42–47; free will and, 61; genes and, 63–65, 100; gray matter, 65–66; identity and, 61–62; ketamine and, 78; lithium and, 77; Low Road, High Road model for emotional processing in, 133–35, 134; memory and, 59, 190; neurobiology of suicide and, xiii–xiv, 57–79, 100–102, 190; pain and, xiv, 42–47, 67–68, 130–31, 133–35, 211–13; PET imaging and, 42; PFC, xiii–xiv, 42–47, 63, 65–66, 68–70, 117–18, 133–35; placebo effect and, 75–76; plasticity of, 59; PTSD and, 41–42; rational, xiii–xiv, 212; serotonin and, 62–65, 78; storytelling and, 117–18; stress and, 37–55, 64–68, 211–12, 275–76; stress hormones and, 69–70, 133–35, 275–76; in suicidal mode, xiii–xiv, 41–47, 49–51, 277; suicide and, 57–79; suicide attempts and, 38–40, 44–45, 47–55; synapses, 59–60; trauma and, 41–42, 64–65, 67

British Journal of Psychiatry, 5–6, 14, 20, 257–58

British Medical Journal, 204

Brodmann Area 46, 45–46, 46, 66

Brown, Gregory K., 173, 182–83, 197, 202

Bruffaerts et al., 132

Buie, Dan, 265

Building a Therapeutic Alliance with the Suicidal Patient (Jobes), 177–78

bullying, 234–35, 245, 246

Cade, John, 77

"Call to Clarify Fuzzy Sets, A" (Berman and Silverman), 249

calming down, 289

CAMS. *See* collaborative assessment and management of suicidality

Caplan, Gerald, 21–22

caring letters, 183–84, 185, 219

case formulation, xv, 12–13, 252–53, 257

Caspi et al., 64

CBT. *See* cognitive behavioral therapy

cerebrospinal fluid, 62–63

change, 21, 51, 59, 68, 95, 117–18, 125

#chatsafe project, 236

Cherkis, Jason, 219

children, 59, 132; attachment, child development, and, 115–17; first learning of suicide as, 188–89; sexual abuse of, 41–42; trauma and, 41–42, 132, 150–51, 192

chronic suicidality, xv

cognition, 50, 58, 144

cognitive behavioral therapy (CBT), 49–50, 96, 120, 172–73; MCBT, 135, 177
cognitive destruction, 132
cognitive myopia, 212
cognitive reframing, 68
"Collaborating to Prevent Suicide" (Jobes), 161–64
collaboration: communication and, 90–91, 161–64, 163, 185–200, 209–10, 263–66; joint case conceptualization and, 191–94; in psychoeducation, 190–91; recommendations, for clinical practice, 267–73; therapeutic relationship and, 161–64, 163, 185–200, 263–66; in video playback, 186–90
collaborative assessment and management of suicidality (CAMS), 120, 170, 254, 263
Commitment to Treatment Statement (CTC), 262
communication: attentive silence and, 148; collaboration and, 90–91, 161–64, 163, 185–200, 209–10, 263–66; connection and, 88–91; listening and, 34, 88–91, 148–59, 166; in medical consultations, 81–91; men and, 83–85, 87–88; problems with, 81–91, 86, 87, 209; recommendations, for clinical practice, 267–73; suicidal thoughts and, 81–83; suicide prevention and, 22–23, 34, 81–82, 87–91, 251–54; suicide risk factors and, 22–23, 209; therapeutic relationship and, 81–91, 86, 87, 121, 148–59, 161–64, 251–54, 263–64; Tower of Babel syndrome and, 85–87, 86, 87, 209; warning signs and, 230–31
confidentiality, 228–29
contagion effect, 24–27, 246–48
Contagious (Berger), 248
controlled breathing, 241
coping mechanisms, 148; controlled breathing as, 241; dysfunctional, 97; online, 241; for pain, 97; for young men, 238–39; for youth, 233–34, 238–39
cortex, 63, 65–70
cortisol, 69
COVID-19, 169, 237–38
crisis: defining, 21–22; emotions in, 22, 211–13; intervention, 21–22, 90–91, 219–20; triggers and, 22
Crisis, 14–15, 73; "An Exercise in Improving Suicide Reporting in Print Media," 26–27; "Short Report" in, 230–32
"Crisis and Suicide" campaign, 21–24
crisis lines, 266, 290; Crisis Text Line Project, 240–41; for listening, 240–41; online, 240–41
CTC. *See* Commitment to Treatment Statement

culture: identity and, 125–26; life goals and, 125–26; social norms and, 103–7, 125–26; suicide and, 103–7, 125–26
Curtis, Cate, 229
cyberbullying, 235, 245, 246

danger, 181; processing, 134, 218
data collection, 29–30, 202, 203–4
DBT. *See* dialectical behavior therapy
deceit, 67
decision making, xiii–xiv; brain and, 65; errors in, 186, 187, 189; SDM, 264
deconstructed mental state, 94–95
De Leo, Diego, 28, 73
depression, 3–4, 7; antidepressant medications and, 71–79, 81–82, 107–8; bipolar disorder and, 77; brain and, 69–79; campaigns for fighting, 70–72; causes of, 11; confusion about, 11, 14; COVID-19 and, 238; depressive episodes and, 11, 75, 139; diagnosis of, 10–12, 14, 17, 69–71, 143–46; Gotland Study on, 17, 21–23, 35, 83; issues in diagnosing, 11–12, 14; ketamine and, 78; men and male depressive syndrome, 83–85; neurobiology and, 69–79; PFC and, 69–70; placebo effect and, 75–76; "reason" for, 11–12; research on, 11–12; stress and, 69–71; suicide in relation to, 70–79, 99; suicide rates and, 74; suicide risk factors and, 11–12, 14–15, 70–75, 99, 276–77; symptoms of, 11; treatment for, 11, 70–79; treatment-resistant, 107–8; in youth, 70
Devitt, Patrick, 257–58
Dey, Pascal, 164
diabetes, 85
Diagnostic and Statistical Manual of Mental Disorders (DSM): classification system and, 13–14; on dissociation, 50–51
dialectical behavior therapy (DBT), 97, 120, 172, 175
diathesis, 100
"Discovering the Truth in Attempted Suicide" (Maltsberger), 169
dissociation, 276; *DSM* on, 50–51; pain and, 48, 51, 94, 146; PDI questionnaire and, 47–51; suicidal mode and, 49–51, 140–42, 146; trauma and, 47–51
Doctor, His Patient, and the Illness, The (Balint), 89
"Don't Be Afraid of Suicidal Patients" workshop, 251–52
DSM. *See Diagnostic and Statistical Manual of Mental Disorders*
dual processing theory, 186, 187–89, 194–95

early warning signs, xiv, 278. *See also* warning signs
ECT. *See* electroconvulsive therapy

EDA. *See* electrodermal activity
electroconvulsive therapy (ECT), 2–3, 11
electrodermal activity (EDA), 37–40, *39*, 44
emergency room visits, 250, 256–57, 270
emotional brain, xiv, 130–31, 133–35
Emotional Brain, The (LeDoux), 133–35
emotions: BPD and, 97; brain and, 58, 62–68, 211–13; calming down, 289; in crisis, 22, 211–13; EDA and measuring, 37–40, *39*, 44; how to talk to person in emotional crisis, 287–88; hypersensitivity and, 66, 97; Low Road, High Road model for emotional processing, 133–35, *134*; men and, 87–88; regulation of, xvi, 62–68, 97, 135, 175; serotonin and, 62–65; shame and, 226–32; in suicidal mode, 41–42, 50; therapy and, 135
empathy, 12–13, 37, 171, 254; mirror neuron, 40
epigenetics, 64–65, 100
EPSIS. *See* European Parasuicide Interview Schedule
Erikson, Erik, 125
escape, 94–95, 130, 141–42, 188
EURO Multicenter Study Group on Suicidal Behavior, 28–30
European Parasuicide Interview Schedule (EPSIS), 29–30, 31, 157

European Symposium on Suicidal Behavior, 103–7
Evolution of Suicide, The (Soper), 130
"Exercise in Improving Suicide Reporting in Print Media, An" (*Crisis*), 26–27
explanatory theories, 13

facial expressions, 66
failure: humiliation and, 125–26; life goals and, 111–14, 124–25; shame and, 125–28; vulnerability and, 111–14
family: interviews, 8–12, 225–27; suicide prevention and, 223–25, 233, 243–44, 287–90; what to do if you're concerned about someone, 287–90
Farberow, Norman, 33, 93–94
Fawcett, Ian, 11–12, 231
fear, 53; of health care professionals, 250–52; of sharing suicidal thoughts, 106, 141, 250–51; vulnerability and, 40–41, 143–46
Firestone, Lisa, 176
Fitzpatrick, Scott J., 216
5-hydroxyindoleacetic acid (5-HIAA), 62, 63–64
Fluid Vulnerability Theory, 259, 260
fMRI. *See* functional magnetic resonance imaging
free will, 250; autonomy and, 206, 218, 288–89; brain and, 61

Freud, Sigmund, 3, 14, 58
Frey, Conrad, 21
functional magnetic resonance imaging (fMRI), 42–47

Gaily-Luoma, Selma, 255
Galynker, Igor, 264–65
gatekeeper training standards, 255
Gelder, Michael, 5–6
General Psychopathology (Jaspers), 12
genes, 63–65, 100
gene x environment interaction, 59, 64
Gilburt, H., 88–89
global suicide rates, xi, xvi
glucocorticoid receptor (GR), 69
Goethe, Johann Wolfgang von, 24, 188–89, 247
Goldblatt, Mark, 172
good clinical practice, 249–50
Gotland Study, 17–18, 21–23, 35, 83
GR. *See* glucocorticoid receptor
gray matter, 65–66
guilt, 211; bereaved parents and, 230; patient suicide and, 3–4; suicide and, 230
gut feeling, 270

Handbook of Narrative and Psychotherapy (Bruner), 118
Hansson, Anna-Lena, 233–34
HAq. *See* Helping Alliance Questionnaire
Harding et al., 103
Haring, Christian, 28
Haugen, Frances, 245

Hawgood, Jacinta, 255
health care cost, 83–84, 204
health care professionals: Aeschi Guidelines for Clinicians, 252–54; fear of, 250–52; gatekeeper training standards for, 255; good clinical practice and, 249–50; gut feeling of, 270; medical consultations and, 81–91; misuse of medical diagnoses and, 13; negative feelings of, 265–66; primary care providers and, 17, 20; recommendations, for clinical practice, 267–73; SDM and, 264; suicide prevention and, 249–73; suicide risk factor assessments and, 257–66, 269–70, 272–73; youth and, 255. *See also* primary care physicians
health literacy: challenges in improving suicide literacy and, 217–18; loss and, 70, 104, 129, 151, 217; National Action Plan to Improve Health Literacy, 215; online, 215; results of suicide literacy and, 216–17; suicide literacy and, 215–20; suicide prevention and, 214–21
Heatherton, T. F., 135, 267
help: acceptance of, 32–33, 104–5, 207–21; seeking, 278
Helping Alliance Questionnaire (HAq), 153–55, 165, 203–4, 272
High Risk for Suicide List, 260–61
Hines, Kevin, 53, 188–89

Hohler, August E., 8
Holmes, Jeremy, 116–17; Aeschi Working Group and, 173–74
home, as place of risk, 262
homework task, 184, 191
hormones, 69–70, 133–35, 275–76
hospital records, 98–99
human condition: fragility of identity and, 123–37; pain and, 96, 130
humiliation, 125–26
Hungary, 125–26
Husserl, Edmund, 12
hypersensitivity, 66, 97

ideation-to-action framework, 258–59
identity: brain and, 61–62; cognitive destruction and, 132; culture and, 125–26; fragility of humans and, 123–37; life goals and, 111–14, 124–28; memory and, 61–62; negative self-beliefs and, 41, 95, 96–97, 111–14, 132, 144–47, 156, 253; pain and, 111–14, 123–37; self-confrontation with narratives and, 148–52, 163–68, 184; shame and, 108, 125–28, 227; suicidal thoughts and, 125–26; vulnerability and, 5, 146–47; youth and, 227, 232–34, 236–39. *See also* professional identity
iGen (Twenge), 244–45
"Imitation Risk Score," 24–27
indigenous populations, 213

influencers, 235–36
Influential Mind, The (Sharot), 158
inpatient care, 22, 261–62
Insel, Thomas, 100–102
interpersonal relations, 30–33, 97; pain and, 131–32
interpersonal theory, 97
intervention: crisis, 21–22, 90–91, 219–20; therapy and, 200–206, 219–20. *See also* Attempted Suicide Short Intervention Program
interviews: conducting, 114–15; with family, 8–12, 225–27; HAq and, 153–55, 165, 203–4, 272; listening during, 148–56, 158–59; narrative interviewing and, xiv–xv, 114–20, 139–56, 163–68, 184–90; with primary care physicians, 9–11; storytelling and, 114–17; with suicidal patient family members, xiii, 8–12; with suicidal patients, xiii; after suicide attempts, 29–34, 136, 139–59; on suicide risk factors, 8, 9–12, 14–15
intrapersonal relations, 30–33, 132
Isacsson, Göran, 71

Jaspers, Karl, 12
Joannidis, John, 204
Jobes, David, 90; Aeschi Working Group and, 166, 170, 177–78; *Building a Therapeutic Alliance with the Suicidal Patient* by, 177–78; CAMS and, 254, 263;

"Collaborating to Prevent Suicide" by, 161–64; SSF and, 263
Joiner, Thomas, 97
joint case conceptualization, 191–94
Jollant, Fabrice, 66
Jung, Carl, 3, 14

Kahneman, Daniel, 186, 189
Kandel, Eric R., 57–59, 78, 117–18, 178, 200
Karolinska Institute, 62–63, 71
Kerkhof, Ad, 28
ketamine, 78
Kirchner, Ludwig, 123–25
Kolakowska, Tamara, 5–6
Krise and Suizid prevention campaign, 21, 23–24
Kuramoto-Crawford et al., 83–84

Lanius, Ruth, 43–44
learning, 57–59
LeDoux, Joseph, 133–35
Leenaars, Antoon A., 164
letters: ASSIP and, 197–200, 203; caring, 183–84, 185, 219; suicide notes and, 93–94, 112; as therapy, 183–85, 197–200, 203, 219
LGBTQ+ community, 236–37
libertarian paternalism, 218
Libet, Benjamin, 60–61
life experience, 30–34, 95–96; in biographical context, 107–21; storytelling and, 114–20
life goals, 121; culture and, 125–26; failure and, 111–14, 124–25; identity and, 111–14, 124–28; professional identity and, 111–14, 124–25; youth and, 223–25
Linehan, Marsha M., 97, 172, 175
listening, 239; attentive silence and, 148; communication and, 34, 88–91, 148–59, 166; crisis lines for, 240–41; importance of, 34, 88–91, 106; during interviews, 148–56, 158–59; in narrative interviews, 148–56; in not-knowing position, 121; storytelling and, 114–20; suicide prevention and, 34, 106, 214; therapeutic relationship and, 114–20, 148–59, 166; vulnerability and, 158–59; workshops, 89–91
literacy, 215–20. *See also* health literacy
Literacy of Suicide Scale (LOSS), 70, 104, 129, 151, 217
lithium, 77
Litman, Robert, 33
Littlemore Hospital, 6–7
LivingWorks ASIST project, 219–20
long-term risk factors, 260–62
Lönnqvist, Jouko, 28
Low Road, High Road model, 133–35, *134*
lying, 30–31, 109, 127, 225–26

machine learning, 258–61
male depressive syndrome, 83–84

Maltsberger, John "Terry," 203, 265; Aeschi Working Group and, 163–64, 165–66, 169, 171–72; "Discovering the Truth in Attempted Suicide" by, 169
mania, 77
manipulation, 30–31, 256
Mann, J. John, 62–63, 100, 249
Maris, Ronald W., 95–96, 107, 113–14
mask, 165, 226–27, 245, 248
MCBT. *See* mindfulness-based cognitive therapy
McCabe, Rose, 261
media: on antidepressant medications, 75; guidelines on suicide reporting for, 25–27; suicide reporting by, 23–27. *See also* social media
medical model: algorithms, machine learning and, 258–61; collaboration compared to, 161–64, 163; communication problems and, 81–91; major issues with, 81–85, 105–7, 209–10, 220–21; mental health and, xii, 81–91, 98–102; one-size-fits-all approach and, xiii, 91, 176, 249; psychological models and, 98–102; suicide-as-illness model and, xii, xvi; suicide is not an illness of, 103–21, 182; suicide risk factors in, xii; view on, 98–102, 157–58, 220–21

memory: brain and, 59, 190; identity and, 61–62; trauma and, 40–41, 46
men: communication and, 83–85, 87–88; coping mechanisms for, 238–39; depression and male depressive syndrome, 83–85; emotions and, 87–88; suicidal thoughts and, 83–85; young men and suicide risk, 127–28, 223–33, 238–39
mental health: COVID-19 and, 237–38; medical model and, xii, 81–91, 98–102; misuse of medical diagnoses and, 13; social media and, 235–36; suicide in relation to, 97, 98–102; support, 289; YAM and, 241–42
mentalization, 148–49
Mentalizing in Clinical Practice (Allen), 175
message, 89; on suicide, 211–13
mindfulness-based cognitive therapy (MCBT), 135, 177
mini-exposure sessions, 185, 197
mirror neuron empathy, 40
misuse of medical diagnoses, 13
modes concept, 96
"Monitoring Study," 28–30
Motto, Jerome A., 5, 183–84, 219
Murphy, George E., 10
Murray, Declan, 257–58
Myers, Michael, 265–66
MYPLAN safety plan, 242

narrative competence, 118, 186
narrative interviewing, xiv–xv,
114–20; Aeschi Working
Group and, 163–68; in ASSIP,
185–86; listening in, 148–56;
overview and experiments in,
139–56; overview and
explanation of, 185–86;
self-confrontation and video
playback in, 149–52, 163–68,
184, 186–90
narrative medicine, 157–58, 256–57
narratives, 61–62; of pain, 129–30;
self-confrontation with, 148–52,
163–68, 184; storytelling and,
114–20, 140–56, 256–57; trauma
and, 118
National Action Plan to Improve
Health Literacy, U.S., 215
National Health Service (NHS), 1
National Institute of Mental
Health (NIMH), 100–101
National Suicide Prevention
Lifeline, 266
Nazi Germany, 123–24
NE. *See* norepinephrine
"Need for Innovation in Health
Care Systems to Improve
Suicide Prevention, The"
(Stanley and Mann), 249
needs, xii, xv–xvi; vulnerability
and, 9–10, 59, 147, 158–59,
192–94, 260; of youth, 233–34
negative feelings, of health care
professionals, 265–66

negative self-beliefs: failure and,
111–14; pain and, 41, 95, 96–97,
111–14, 132, 144–47, 156, 253;
unbearable mental state and,
156, 192–93, 253
Netflix, 246–48
neurobiology: behavior and, 57–58,
130–31; Cajal and, 59–60;
depression and, 69–79; EDA
and, 37–40, 39, 44; Kandel on,
57–59; Libet and, 60–61;
neurons, synapses, and
neurotransmitters, 59–60;
serotonin and, 62–65; of suicide,
xiii–xiv, 57–79, 100–102, 190
"New Intellectual Framework for
Psychiatry, A" (Kandel), 57–58
NHS. *See* National Health
Service
NIMH. *See* National Institute of
Mental Health
no-problem mask, 226–27
norepinephrine (NE), 63

obsessive-compulsive disorder
(OCD), 108–9
Ochsner, Kevin, 135
one-size-fits-all approach, xiii, 91,
176, 249
online resources: health literacy,
215; for suicide prevention,
240–43; for youth, 240–43
on/off phenomenon, xiii–xiv, 50.
See also dissociation
opioid dependency, 124

Orbach, Israel, 51, 94, 129; Aeschi Working Group and, 164, 171
outreach elements, 183–84
overdose, 10, 29, 30–33, 49, 193–94, 256

pain: brain and, xiv, 42–47, 67–68, 130–31, 133–35, 211–13; coping mechanisms for, 97; dissociation and, 48, 51, 94, 146; human condition and, 96, 130; humiliation and, 125–26; identity and, 111–14, 123–37; interpersonal relations and, 131–32; narratives of, 129–30; negative self-beliefs and, 41, 95, 96–97, 111–14, 132, 144–47, 156, 253; psychache and, 94, 128–33; questionnaire, 94; self-extinction and, 130; shame and, 108, 125–28, 227–33; soul, 48; strength of, 129–30; stress and, 67–68, 128–33; unbearable mental state and, 48, 52, 94–97, 111–14, 128–33, 156, 195–96, 212, 253, 275; vulnerability and, 67–68, 121, 139–59
parental support: bereaved parents and, 126–28, 224–26, 230–32; what to do if you're concerned about someone, 287–90; youth and, 126–28, 223–25, 233, 243–44
paroxetine, 72
paternalism, 218
Pathways to Suicide (Maris), 95–96
patient handout, ASSIP, 275–79

patient suicide, xi; experiencing, 3–6; guilt and, 3–4; inpatient, 22, 261–62; insights from, 9–10; interviews, with suicidal patient family members, 8–12; losing a patient and, 1–15; personal characteristics in, 8–9; psychiatric disorders and, 1–7; research on, 5–12; at Roffey Park Hospital, 1–5; surprise over, 10; therapeutic relationship and, 3–4, 88–91
Peritraumatic Dissociation Index (PDI), 47–51
personal experiences: in biographical context, 107–21; life experience, suicide, and, 30–34, 95–96, 107
personal values and ethics, 109
PET. *See* positron emission tomography
Peterson et al., 263–64
PFC. *See* prefrontal cortex
pharmaceutical industry, 71–74
phenomenology, 12–13
"Physician's Responsibility for Suicide" (Murphy), 10
Pinfold, Terence, 85–86
placebo effect, 75–76
plasticity, 59
Platt, Stephen, 28
PLOS Medicine, 201, 203
politics, 13
Pollock, L. R., 95
positron emission tomography (PET), 42

post-traumatic stress disorder
(PTSD), 1; brain and, 41–42;
veterans and, 81–82
prefrontal cortex (PFC), xiii–xiv;
amygdala and, 43, 43–44,
68–69, 133–35; Brodmann
Area 46, 45–46, 46, 66; cortex
and, 65–66; depression and,
69–70; imaging, 42–47;
serotonin in, 63; storytelling
and, 117–18
Present State Examination, 114
primary care physicians:
communication problems with,
81–91, 86, 87; "Crisis and
Suicide" campaign for training,
21–24; gatekeeper training
standards for, 255; interviews
with, 9–11; listening workshops
for, 89–91; suicidal thoughts
and, 19–20; suicide prevention
and, 17–19
primary care providers, 17, 20
Principles of Preventive Psychiatry
(Caplan), 21–22
problem-solving skills, 97, 212
professional identity, xii, 266;
failure and, 111–14; life goals
and, 111–14, 124–25; personal
life and, 224–25; vulnerability
and, 5
pseudodementia, 145–46
psychache, 94, 128–33
psyche, 93
psychiatric disorders: changes in
classifying, 14; classification of,
13–14; dangers of explanatory
theories of, 13; *DSM* and,
13–14; misuse of medical
diagnoses and, 13; patient
suicide and, 1–7; research on,
6–8; suicide in relation to, 1–8,
97, 98–102; suicide prevention
and, 7–8. *See also specific
disorders*
psychological autopsy, 98–99
psychological models: choice of,
93–98; medical model and,
98–102
psychological pain. *See* pain
Psychology Today, 101–2
PTSD. *See* post-traumatic stress
disorder

QPR Gatekeeper Training for
Suicide Prevention, 20

Ramon y Cajal, Santiago,
59–60
randomized controlled studies,
200–205
Rasmussen, Mette, 127–28
rational brain, xiii–xiv, 212
Rational Suicide, Irrational Laws
(Stefan), 105–6
"Real Men, Real Depression"
campaign, 83
reason-cause discussion, 156
regret, 53, 212
Reisch, Thomas, 43–44
"Repetition-Prediction Study,"
28–30, 31

research: ASSIP randomized controlled study and, 200–205; on depression, 11–12; Gotland Study and, 17–18, 21–23, 35, 83; pain questionnaire and, 94; on patient suicide, 5–12; PDI questionnaire and, 47–51; on psychiatric disorders, 6–8; psychological models and, 93–98; questionnaire, for suicide prevention, 17–20; on suicidal behavior, 28–33, 103–7; on suicide risk factors, 8–12, 14–15, 100–102, 173; using psychological autopsy, 98–99. *See also* interviews

resources, 291–95; online, 215, 240–43; therapeutic, 182

right to suicide, 7, 18, 104, 218

Robinson, Jo, 255

Roffey Park Hospital, 1–5

Rogers, Jim, 90–91, 176

Ronningstam, Elsa, 172

Royal College of Psychiatrists, 6, 12

Rudd, David, 41, 49–50, 96, 202; Aeschi Working Group and, 172; CTC and, 262; Fluid Vulnerability Theory and, 259

Rutz, Wolfgang, 17, 83. *See also* Gotland Study

safeTALK, 220

safety planning, 185, 278; in ASSIP, 194–97, *196*; MYPLAN safety plan, 242; Suicide Safety Plan app, 240

safety strategies, 50, 185, 194, 196–97, 267, 277, 284

Saving and Empowering Young Lives, 241–43

Schechter, Mark, 172

schizophrenia, 13

Schmidtke, Armin, 28

school, 238

Schwab, Simon, 202, 203–4

Scientific American, 257, 258

Scotland, 27–28

SCS. *See* Suicide Crisis Syndrome

SDM. *See* Shared Decision Making

Search for the Secure Base, The (Holmes), 116–17

seeking help, 278

Sekasin, 240

Self-Assessed Expectations of Suicide Risk Scale, 263–64

self-confrontation, xiv–xv, 148–52, 163–68, 184

self-criticism. *See* negative self-beliefs

self-esteem, 66, 111–14, 146–47

self-extinction, 130

self-harm, 41, 43–44, 48, 73, 96–97, 143–45, 256–57

self-soothing, 289

sensationalism, 25–27

separation anxiety, 38–40, 132, 139–43, 149–52

serotonin: antidepressant medications and, 72; brain and, 62–65, 78; emotions and, 62–65

sexual abuse, 41–42

shame, 211; defining, 227; failure and, 125–28; humiliation and, 125–26; identity and, 108, 125–28, 227; pain and, 108, 125–28, 227–33; self-criticism and, 41; suicide and, 108, 125–30, 226–32; talking about, 228–30; understanding, 227–29; youth and, 126–28, 226–32, 234–35

Shared Decision Making (SDM), 264

Sharot, Tali, 158

Shneidman, Edwin, 33–34, 93–94, 129, 164, 165

"Short Report" (*Crisis*), 230–32

short-term risk factors, 260–63

snail (*Aplysia californica*), 57–58

social contagions, 24–27, 246–48

social media: addiction to, 245; #chatsafe project and, 236; cyberbullying and, 235, 245, 246; influencers, 235–36; mental health and, 235–36; suicide prevention and, 240–43; suicide risk factors and, 235–36, 246, 248; young girls and, 233, 244–45; youth and, 235–36, 241–48; YouTube and, 242–43

social norms, 103–7, 125–26

Sonneck, Gernot, 26–27

Soper, C. A., 130–31

Sorrows of Young Werther, The (Goethe), 24, 188–89, 247

soul pain, 48

South Korea, 125

Soviet Union, 13

Soyka, Karen, 90–91

SSF. *See* Suicide Status Form

Stanley, B., 249

Stefan, Susan, 105–6

stigma, 218–19

storytelling: brain and, 117–18; joint case conceptualization and, 191–94; narratives and, 114–20, 140–56, 256–57; therapeutic relationship and, 114–20, 256–57

stress: brain and, 37–55, 64–68, 211–12, 275–76; depression and, 69–71; genes and, 64–65, 100; hormones, 69–70, 133–35, 275–76; overactive, 100; pain and, 67–68, 128–33; PTSD and, 1, 41–42, 81–82; from separation anxiety, 38–40, 132, 139–43, 149–52; suicide risk factors and, 276; trauma and, 38–42, 64–65, 67, 276

stress-diathesis model, 100

"Stubborn Behaviour, A" (Van Praag), 72

substance abuse: alcohol and, 277; ketamine and risks of, 78; opioid dependency and, 124; overdose and, 10, 29, 30–33, 49, 192–94, 256; suicidal thoughts and, 277; suicide attempts and, 30–33, 45, 49, 193–94, 256; suicide risk factors and, 15, 29, 124, 277 suicidal behavior: antidepressant medications and increase in, 73–75; biographical context and, 107–21; calming down, 289; data collection on, 29–30; early warning signs and, xiv, 278; EDA for detecting, 37–40, 39, 44; European Symposium on Suicidal Behavior, 103–7; help acceptance and, 32–33, 104–5, 207–21; how to talk to person in emotional crisis, 287–88; monitoring, 28–30; new models for framing, 254; past history of, 253; personal experiences of, 30–33; physician opinions *versus* patient opinions of, 30–33, 104–5; recognizing, 21–23; research on, 28–33, 103–7; suicidal thoughts and, 19–20, 105–6; suicide attempts and, 30–33, 44–45, 47–49, 139–43; Suicide Crisis Syndrome, 265; vulnerability and, 139–59; what not to say during, 288; what you must know about, 214; WHO/EURO Multicenter Study Group on Suicidal Behavior and, 28–30

suicidal mind, xv–xvi, 30–34, 95–98, 110–11, 135–36, 218–19

suicidal mode: bodily functions in, 50; brain in, xiii–xiv, 41–47, 49–51, 277; cognition in, 50; description of, 41–42, 49–50; dissociation and, 49–51, 140–42, 146; emotions in, 41–42, 50; insights from, 53–54; modes concept and, 96; suicide attempts and, 277; suicide prevention through understanding, 50; switching "off," 53–54; triggers for, 50, 52, 140–43, 258; unbearable mental state and, 41, 46, 52, 195–96; warning signs of, 50, 211–13

suicidal thoughts: asking about, 239; communication and, 81–83; fear and risk of sharing, 106, 141, 250–51; identity and, 125–26; increased medical visits related to, 19–20; as kept private or secret, 19–20, 84, 105–6; men and, 83–85; primary care physicians and, 19–20; rates of, 105–6; Self-Assessed Expectations of Suicide Risk Scale and, 263–64; self-awareness about, 212–13; substance abuse and, 277; suicidal behavior and, 19–20, 105–6; what to do when you're having, 278; what you must know about, 214; in youth, 105

suicide: action, xiii; as action, not illness, 103–21, 182; box theories explaining, 93–102; brain and, 57–79; culture and, 103–7, 125–26; depression in relation to, 70–79, 99; first introduction to concept of, 188–89; guilt and, 230; learning more about, 289;

literacy, 215–20; mental health in relation to, 97, 98–102; message to share about, 211–13; neurobiology of, xiii–xiv, 57–79, 100–102, 190; psychiatric disorders in relation to, 1–8, 97, 98–102; right to, 7, 18, 104, 218; shame and, 108, 125–30, 226–32; story, xiii; "ten commonalities" of, 94; understanding reasons for, 30–33, 106, 128, 135–36, 225–27. *See also specific topics*
"Suicide, from Genes to Molecules to Treatment and Prevention" (Insel), 100–101
suicide-as-action framework, xiii, xv, 157–59, 218–21, 267
"Suicide as Escape from Self" (Baumeister), 94–95
suicide-as-illness model, xii, xvi. *See also* medical model
"Suicide as Psychache" (Shneidman), 94, 129
suicide attempts: Aeschi Working Group on understanding, 162–79; brain and, 38–40, 44–45, 47–55; EPSIS and, 29–30, 31, 157; follow-up treatment and, 209–10; immediate regret after, 53, 212; interviews after, 29–34, 136, 139–59; past history and, 253; per year, xvi–xvii; positive outcomes after, 32–33; reasons listed for, 30–33; stories of, 38–40, 44–45, 47–55, 136, 139–54, 156–57, 192–94; substance abuse and, 30–33, 45, 49, 193–94, 256; suicidal behavior and, 30–33, 44–45, 47–49, 139–43; suicidal mode and, 277; therapy after, xiv–xv, 209–10; youth, 224, 230–34, 242. *See also* Attempted Suicide Short Intervention Program
Suicide Call Back Service, 240
suicide careers, 113–14
Suicide Crisis Syndrome (SCS), 265
"Suicide Is Not a Rational Act" handout, 191
suicide notes, 93–94, 112
suicide prediction, 28–31, 257–64, 269–70
suicide prevention: advice giving and, 288; Aeschi Guidelines for Clinicians and, 252–54; antidepressant medications and, 71–79, 81–82; ASSIP patient handout and, 275–79; attitude toward, 288–89; availability and limits of, 288; awareness for, 241–43; caring letters and, 183–84, 185, 219; challenges of, xvi; communication and, 22–23, 34, 81–82, 87–91, 251–54; "Crisis and Suicide" campaign for, 21–24; *Crisis* letter on, 230–32; CTC and, 262; early treatment goals and, 270–72; emergency room visits and, 250, 256–57, 270; family and, 223–25, 233, 243–44, 287–90; first lessons in,

suicide prevention (*continued*)
17–35; first remedy for, 214; Gotland Study and, 17–18, 21–23, 35, 83; health care professionals and, 249–73; health literacy and, 214–21; help acceptance and, 32–33, 104–5, 207–21; how to talk about suicide and, 216; how to talk to person in emotional crisis, 287–88; innovation needs for, 249–73; with ketamine, 78; Krise and Suizid prevention campaign, 21, 23–24; listening and, 34, 106, 214; with lithium, 77; LivingWorks ASIST project and, 219–20; new therapeutic model for, 182–206; online resources for, 240–43; outreach elements, 183–84; with person-centered approach, 210–21; primary care physicians and, 17–19; psychiatric disorder and, 7–8; QPR Gatekeeper Training for Suicide Prevention and, 20; questionnaire, 17–20; recommendations, for clinical practice, 267–73; safeTALK and, 220; safety planning and, 185, 194–97, 196, 240, 242, 278; social media and, 240–43; in Switzerland, 17–18, 21, 27–30, 266; therapeutic relationship and, 22–23, 34, 88–91, 271; treatment goals and, 182–83, 252–54, 267, 269; understanding suicidal mode for, 50; what not to say and, 288; what to do if you're concerned about somebody, 287–90; WHO and, 27–30; for youth, 230–34, 238–44, 255

suicide rates, 213; depression and, 74; in England, 6–7; global, xi, xvi; Gotland Study and decrease in, 17–18; huge differences in, between countries, 27–30; LGBTQ+, 236–37; media reporting and, 23–27; per year, xvi–xvii; in Switzerland, 6–7, 24–25, 28–29; youth, xvi–xvii, 244, 246–47

suicide reporting: guidelines for, 25–27; by media, 23–27

"Suicide Risk Assessment Doesn't Work" (Murray and Devitt), 257–58

suicide risk factors: assessments for, 8–9, 88, 250–51, 257–66, 269–70, 272–73; ASSIP on, 275–79; Brown and, 173; communication and, 22–23, 209; contagion effect and, 24–27, 246–48; depression and, 11–12, 14–15, 70–75, 99, 276–77; early warning signs and, 278; Fluid Vulnerability Theory and, 259, 260; High Risk for Suicide List and, 260–61; home and, 262; ideation-to-action framework on, 258–59; "Imitation Risk Score" and,

24–27; inpatient, 22, 261–62; interviews on, 8, 9–12, 14–15; issues in understanding, 15; LGBTQ+, 236–37; long-term *versus* short-term, 260–63; in medical model, xii; predictions on, 28–31, 257–64, 269–70; research on, 8–12, 14–15, 100–102, 173; Self-Assessed Expectations of Suicide Risk Scale and, 263–64; social media and, 235–36, 246, 248; SSF and, 263; stress and, 276; stress-diathesis model and, 100; substance abuse and, 15, 29, 124, 277; Suicide Crisis Syndrome and, 265; Therapist Response Questionnaire Suicide Form and, 265; warning signs and, 207–8, 230–31, 278; for young girls, 233, 244–45; for young men, 127–28, 223–33, 238–39; youth and, 223–48

Suicide Safety Plan app, 240

Suicide Status Form (SSF), 263

Suizid and Krise prevention campaign, 21, 23–24

Sunstein, C. R., 206, 218

Swiss Medical Association, 21

Swiss Medical Journal, 21

Swiss Medical Weekly, 14

Switzerland, 103, 123–24, 159, 161; Aeschi Working Group and, 164–79; suicide prevention in, 17–18, 21, 27–30, 266; suicide rates in, 6–7, 24–25, 28–29

teaching, 20; collaborative psychoeducation and, 190–91; with "Crisis and Suicide" campaign, 21–24; health literacy, 214–21; listening skills, 89–91

team counseling, 5

Teismann, Tobias, 262–63

terminal illness, 213

Thaler, R. H., 206, 218

therapeutic program: components of, 184–200; designing new, 182–84; intervention and, 200–206, 219–20. *See also* Attempted Suicide Short Intervention Program

therapeutic relationship, xii–xiii; Aeschi Guidelines for Clinicians and, 252–54; Aeschi Working Group and, 162–79; attachment theory and, 115–17, 174; attentive silence and, 148; collaboration and, 161–64, *163*, 185–200, 263–66; communication and, 81–91, 86, 87, 121, 148–59, 161–64, 251–54, 263–64; CTC and, 262; importance of, 34, 76, 88–91, 166; listening and, 114–20, 148–59, 166; for long-term, 271; patient suicide and, 3–4, 88–91; recommendations, for clinical practice and, 267–73; self-confrontation and video playback for, 149–52, 163–68, 184, 186–90; storytelling and, 114–20, 256–57; suicide prevention and, 22–23, 34, 88–91, 271; VTAS and, 76

therapeutic resources, 182
Therapist Response Questionnaire Suicide Form, 265
therapy, xii–xiii; Aeschi Working Group on, 164–79; anonymous counselors and, 228; CAMS, 120, 170, 254, 263; CBT, 49–50, 96, 120, 172–73; change and, 117–18; CTC and, 262; DBT, 97, 120, 172, 175; early treatment goals, 270–72; emotions and, 135; established types of, 119–20; follow-up treatment, 209–10; goals of, 120, 252–54, 267, 269; golden rule of, 185; intervention and, 200–206, 219–20; letters as, 183–85, 197–200, 203, 219; MCBT, 135, 177; mini-exposure sessions, 185, 197; narrative approach to, 114–15, 118–20, 256–57; personal experiences and, 33–34; recommendations, for clinical practice, 267–73; after suicide attempts, xiv–xv, 209–10; team counseling and, 5; therapist think tanks, 161–79; three-session, 200–206; translating acquired knowledge into, 181–206. *See also* Attempted Suicide Short Intervention Program
Thinking Fast and Slow (Kahneman), 186
13 Reasons Why (Netflix series), 246–48
three-session therapy, 200–206

Titanic suicide scene, 88
Todd, Amanda, 244–45
Törnblom, Annelie, 226–27
Tower of Babel syndrome, 85–87, 86, 87, 209
Träskman, Lil, 62
trauma, 9–10; from abandonment, 67, 235; adversity and, 132; brain and, 41–42, 64–65, 67; children and, 41–42, 132, 150–51, 192; dissociation and, 47–51; EDA and, 38–40; genes and, 64–65; memory and, 40–41, 46; narratives and, 118; PDI and, 47–51; PTSD and, 1, 41–42, 81–82; separation anxiety and, 38–40, 132, 139–43, 149–52; stress and, 38–42, 64–65, 67, 276; triggers and, 132–33, 140–43, 146–47, 151–52; vulnerability and, 149–52, 192–94; youth and, 234–35, 246–47
Treloar, Adrian, 85–86
triggers, 67; crisis and, 22; EDA on, 40; for suicidal mode, 50, 52, 140–43, 258; trauma and, 132–33, 140–43, 146–47, 151–52
trust, 228–29, 268
Twenge, Jean M., 244–45

unbearable mental state, xiii–xiv, 30, 32–34; failure and, 111–14; negative self-beliefs and, 156, 192–93, 253; pain and, 48, 52, 94–97, 111–14, 128–33, 156, 195–96, 212, 253, 275; suicidal

mode and, 41, 46, 52, 195–96; in terminal illness and hopeless conditions, 213
unconscious, 93
understanding: Aeschi Working Group on suicide attempts, 162–79; in biographical context, 107–21; in phenomenology, 12–13; reasons for suicide, 30–33, 106, 128, 135–36, 225–27; shame, 227–29; suicidal mind, 30–34, 110–11, 135–36
"Understanding Attempted Suicide" meeting, 164–69, 167

VA. *See* Veterans Administration
Valach, Ladislav, xii–xiii, 37–38, 110–14, 148–49, 152–53, 161, 183; Aeschi Working Group and, 164–67, 176–77
values, 109
Vanderbilt Therapeutic Alliance Scale (VTAS), 76
Van Praag, Herman M., 72–73
veterans, 81–82
Veterans Administration (VA), 81–82, 260–61
video playback, xiv, 149–52, 163–68, 184, 186–90
Virtual Hope Box, 241
VTAS. *See* Vanderbilt Therapeutic Alliance Scale
vulnerability, xv; attachment theory and, 115–17; defining, 67; failure and, 111–14; fear and, 40–41, 143–46; Fluid Vulnerability Theory and, 259, 260; fragility and, 123–37; identity and, 5, 146–47; listening and, 158–59; needs and, 9–10, 59, 147, 158–59, 192–94, 260; pain and, 67–68, 121, 139–59; professional identity and, 5; storytelling and, 114–20; suicidal behavior and, 139–59; trauma and, 149–52, 192–94; triggers and, 67

warning signs: ASSIP, 194–95; communication and, 230–31; early, xiv, 278; lack of, xvi, 225, 230–31; learning more about, 225, 289; of suicidal mode, 50, 211–13; suicide risk factors and, 207–8, 230–31, 278
Wasserman, Danuta, 28
Wegner, Daniel M., 61
Weinberg, Igor, 172
Werther effect, 24–25, 247–48
what not to say, 288
what to do if you're concerned about somebody, 287–90
what to do when you're having suicidal thoughts, 278
what you must know about suicidal thoughts, 214
WHO. *See* World Health Organization
Williams, Mark G., 95
Working Alliance Inventory, Short Form, 272

World Health Organization (WHO), 27–30; on depression, 74; EURO Multicenter Study Group on Suicidal Behavior and, 28–30

YAM. *See* Youth Aware of Mental Health Program
Young, Richard, 149, 176–77
youth: AdoASSIP for, 205, 233; asking about suicidal thoughts, 239; bullying and, 235, 245, 246; contagion effect and, 246–48; coping mechanisms for, 233–34, 238–39; COVID-19 and, 237–38; depression in, 70; health care professionals and, 255; identity and, 227, 232–34, 236–39; lack of warning signs in, 225; LGBTQ+, 236–37; life goals and, 223–25; needs of, 233–34; online resources for, 240–43; parental support and, 126–28, 223–25, 233, 243–44; Saving and Empowering Young Lives program for, 241–43; school and, 238; shame and, 126–28, 226–32, 234–35; social media and, 235–36, 241–48; suicidal thoughts in, 105; suicide attempts, 224, 230–34, 242; suicide prevention for, 230–34, 238–44, 255; suicide rates, xvi–xvii, 244, 246–47; suicide risk factors and, 223–48; summary on, 248; *13 Reasons Why* and, 246–48; trauma and, 234–35, 246–47; Werther Effect and, 247–48; young girls and suicide, 233, 244–45; young men and suicide, 127–28, 223–33, 238–39

Youth Aware of Mental Health Program (YAM), 241–42
Youth-Nominated Support Team Intervention for Suicidal Adolescents- Version II (YST-II), 244
YouTube, 242–43
YST-II. *See* Youth-Nominated Support Team Intervention for Suicidal Adolescents- Version II

Zero Reasons Why, 240

Milton Keynes UK
Ingram Content Group UK Ltd.
UKHW010956121223
434188UK00004B/15/J